Make: Getting Started with Intel Edison

Stephanie Moyerman

Make: Getting Started with Intel Edison

by Stephanie Moyerman

Printed in the United States of America.

Published by Maker Media, Inc., 1160 Battery Street East, Suite 125, San Francisco, CA 94111.

Maker Media books may be purchased for educational, business, or sales promotional use. Online editions are also available for most titles (*http://safaribooksonline.com*). For more information, contact O'Reilly Media's institutional sales department: 800-998-9938 or *corporate@oreilly.com*.

Editor: Roger Stewart
Production Editor: Nicole Shelby
Copyeditor: Gillian McGarvey
Proofreader: Amanda Kersey
Indexer: Ellen Troutman
Interior Designer: David Futato
Cover Designer: Brian Jepson
Illustrator: Rebecca Demarest

November 2015: First Edition

Revision History for the First Edition

2015-11-02: First Release

See *http://oreilly.com/catalog/errata.csp?isbn=9781457187599* for release details.

978-1-457-18759-9

[LSI]

Contents

Preface ... **vii**

1/Introduction to Edison ... **1**
- Tour of Breakout Boards ... **2**
  - The Arduino Breakout Board ... **3**
  - Other Breakout Options ... **6**
- Setup and Configuration ... **11**
- Install ... **13**
  - Mac and Windows ... **14**
- Connecting ... **16**
  - Windows ... **16**
  - Mac ... **19**
  - Linux ... **19**
  - Logging In ... **20**
  - Configuring Edison and Getting Online ... **20**
- Troubleshooting ... **23**
- Going Further ... **24**

2/Introduction to Linux ... **25**
- What Is Linux? ... **25**
- The Edison Filesystem ... **26**
- Basic Linux Commands ... **29**
  - Accounts, Permissions, and Ownership ... **35**
- Scripting and More Advanced Linux Commands ... **38**
- The Internet ... **41**
- Installing Packages in Yocto ... **44**
- Text Editors ... **46**
- Going Further ... **47**

3/Introduction to Arduino ... **49**
- What is Arduino ... **49**
- Materials List ... **50**
- The Arduino IDE ... **52**

Installing..... 52
Navigating the IDE..... 52
Sketches and Functions..... 55
Digital Output with Blink..... 57
Going Further with Blink..... 59
The Blink Circuit..... 61
Digital Input: Adding a Button..... 68
The Serial Console..... 70
Toggling the LED..... 73
Analog Output..... 75
Analog Input..... 78
I2C Accelerometer..... 82
SPI Screen..... 87
Linux, C++, and the Arduino IDE..... 92
Troubleshooting..... 95
Going Further..... 95

4/Programming in Python..... 97
Introduction..... 97
"Hello, World" in Python..... 98
Functions and Loops..... 101
Installing Libraries..... 102
Blink in Python..... 103
Scripting at Bootup..... 105
Button-Controlled Blink..... 106
Bluetooth-Controlled LED..... 106
Bluetooth Pairing..... 107
Exchanging Information..... 109
The Potentiometer..... 112
I2C Accelerometer..... 113
Installing the Dependencies..... 115
Using the MMA Library..... 115
SPI Screen..... 118
BLE Beacon..... 120
Edison Side..... 120
Smartphone Side..... 122
Going Further..... 123

5/Teach Edison to See..... 125
Introduction..... 125
Materials List..... 126

OpenCV. 127
Extracting Colored Objects. 127
Viewing Images. 128
Face Detection. 130
Webcam. 133
Snapping Photos. 134
Recording Video. 135
Streaming Video. 137
Processed Streaming Video. 140
Going Further. 141

6/Exploring Sound. 143
Introduction. 143
Materials List. 143
Connecting a Headset. 144
Playing and Recording Sounds. 145
Makeshift MP3 Player. 146
Recording Audio with Python. 148
Basic Recording. 148
Thresholding. 150
Speech Recognition. 152
Controlling Devices. 154
Going Further. 156

7/Conclusions. 159
Linux Flavors. 159
Programming Languages. 161
Node.js. 161
C and C++. 162
The Intel XDK IoT Edition. 164
Shutdown Now. 165

A/Materials. 167
Glossary. 171

Index. 179

Preface

Intel Edison is ... well, it's hard to say. Intel Edison is so many different things. If someone told you about Edison 10 years ago, you would have thought that person was utterly insane.

At its core, Intel Edison is a very powerful and cheap little computer. Priced at around $50 USD, this dual-core Atom processor is comparable to an entry-level Windows 8 tablet. At only a small fraction of the cost, it contains all the software amenities of modern-day computing. The system contains 1 GB of memory and 4 GB of storage to handle heavy computational tasks and data logging.

Intel Edison is a communications device. It comes integrated with Bluetooth and WiFi capabilities and preinstalled software to run both.

Intel Edison is a hardware development platform. It contains an Intel Quark microcontroller that allows you to program and control connected devices and circuits. It's also Arduino-compatible, meaning the vast majority of shields, code examples, and libraries that have been specifically designed for Arduino will work as is with Edison, too.

Finally, Intel Edison is an embedded device, designed for the *Internet of Things (IoT)* and wearable technology. Even though Intel Edison is a computer, communications device, and hardware development platform, it weighs only eight grams, is approximately the size of a postage stamp, and can run in low power modes.

While on the subject, it's worth taking a moment to discuss what Edison is *not*. Although Edison is a standalone computer running a full Linux operating system, it's meant to be the brains of your connected and wearable devices—not a laptop or desktop in the traditional sense. For this reason, Edison does not interface with a display, keyboard, or mouse on its own; you connect to Edison through a host computer and load your software directly on the module. Edison is not a great choice if you're looking to build a home media console or old-school arcade game (though both could still be done, with a lot of work), but it's a great candidate for a personal assistant, smart watch, automated robot, smart-home controller, or basically any other electronic system you can imagine. What will you make?

Total Cost

The Edison module alone costs around $50, and if you add the accompanying components to complete every exercise in this book, the total cost is around $200. For a complete materials list, see Appendix A.

What You Can Do with It

As you can see, Intel Edison is a versatile platform that can be used for a great many applications:

Learn about computers
: Edison is a great and inexpensive way to learn more about computers at the lowest level. With Edison, you can learn Linux, configure your operating system, learn about kernels and builds, and install and program drivers for your accessories. You can learn about WiFi and Bluetooth networking by programming Edison to be a dynamic wireless hotspot or Bluetooth beacon. Chapter 2 provides an introduction to Linux and the specific operating system (OS) on Edison. Chapter 3 and Chapter 4 touch on Bluetooth commands.

Learn to program
: Edison is an amazing tool for learning to program; it comes preloaded with many different compilers and interpreters,

and installing more is a breeze. From Chapter 4 onward, this book focuses mainly on programming in Python, but Edison also supports Java, node.js, C, C++, and many more languages. Even Arduino programming is expanded with Edison. Standard Arduino programs consist of compiled C++ based on the *avrlibc* library, but Edison exposes the Arduino IDE to the full C++ standard programming environment. With these additional libraries, you can make system calls and tap into the power of Linux within your Arduino sketches, making Arduino for Edison a powerful tool. You'll see how in Chapter 3.

Make
: You can use Edison to program your electronics projects. Edison and its accompanying ecosystem support the same inputs and outputs (I/Os) as the Arduino Uno and integrate seamlessly with almost any existing Arduino project (more on this in Chapter 1 and Chapter 3). With its size, connectivity, and computing power, you can build elaborate and computationally heavy systems that interface directly with the web or store data on board. In Chapter 3, you'll create a range of electronic circuits powered by Edison and program them using the Arduino IDE. Because programming electronics on Edison is not limited to the Arduino IDE, you recreate these same projects in Chapter 4 by programming them in Python. After that, you'll interface some of these electronics with other computational tasks, leveraging the full power of Edison.

Go to market
: Edison was created specifically to lower the barrier to entry for makers hoping to take their ideas and prototypes all the way to product. Edison interfaces with a variety of breakout boards (discussed in Chapter 1), allowing you to rapidly prototype on one while resting assured that your final design will work on another. In this way, you can develop and program without worrying about the final form factor; Edison will handle this for you.

What's Been Done with It

Sometimes it's hard to decide what to do with such an interesting new device, especially one with such a range of options. Although Edison has been, at the time of this writing, in production for under a year, a wealth of amazing projects have already been created and you can look to for inspiration:

- Intel's project gallery (*http://bit.ly/intel-gallery*)
- Hackster's gallery (*https://www.hackster.io/intel-edison*)
- Intel's Make It Wearable finalists (*http://bit.ly/intel-nixie*)
- Hackaday.io (*https://hackaday.io/search?term=intel+edison*)

Who This Book Is For

This book is an introductory tutorial for Intel Edison. It is meant to showcase the versatility of the product and therefore spans a wide range of topics, from Linux to hardware interfaces to Python programming. You don't need experience in any of these topics to get started, just a little curiosity and a desire to build new things. The only thing you'll need are some basic computer skills: the ability to move and manipulate files, search within your computer, and install software.

The aim of this book is to help you get started designing, building, and programming end-to-end systems with Edison. This book is in no way a comprehensive guide to systems engineering or computer science—there are many resources available should you want to delve deeper into either of those topics. Instead, this book is meant to inspire you, so that you can take your ideas and concepts very quickly to reality.

Conventions Used in This Book

The following typographical conventions are used in this book:

Italic
: Indicates new terms, URLs, email addresses, filenames, and file extensions.

`Constant width`
: Used for program listings, as well as within paragraphs to refer to program elements such as variable or function names, databases, data types, environment variables, statements, and keywords.

`Constant width bold`
: Shows commands or other text that should be typed literally by the user.

`Constant width italic`
: Shows text that should be replaced with user-supplied values or by values determined by context.

This element signifies a tip, suggestion, or a general note.

This element indicates a warning or caution.

Using Code Examples

This book is here to help you get your job done. In general, you may use the code in this book in your programs and documentation. You do not need to contact us for permission unless you're reproducing a significant portion of the code. For example, writing a program that uses several chunks of code from this book does not require permission. Selling or distributing a CD-ROM of examples from MAKE books does require permission. Answering a question by citing this book and quoting example code does not require permission. Incorporating a significant amount of example code from this book into your product's documentation does require permission.

We appreciate, but do not require, attribution. An attribution usually includes the title, author, publisher, and ISBN. For example: *Make: Getting Started with Intel Edison* by Stephanie Moyerman (Maker Media). Copyright 2016, 978-1-4571-8759-9.

If you feel your use of code examples falls outside fair use or the permission given here, feel free to contact us at *bookpermissions@makermedia.com*.

Safari® Books Online

Safari Books Online is an on-demand digital library that delivers expert content in both book and video form from the world's leading authors in technology and business.

Technology professionals, software developers, web designers, and business and creative professionals use Safari Books Online as their primary resource for research, problem solving, learning, and certification training.

Safari Books Online offers a range of plans and pricing for enterprise, government, education, and individuals.

Members have access to thousands of books, training videos, and prepublication manuscripts in one fully searchable database from publishers like O'Reilly Media, Prentice Hall Professional, Addison-Wesley Professional, Microsoft Press, Sams, Que, Peachpit Press, Focal Press, Cisco Press, John Wiley & Sons, Syngress, Morgan Kaufmann, IBM Redbooks, Packt, Adobe Press, FT Press, Apress, Manning, New Riders, McGraw-Hill, Jones & Bartlett, Course Technology, and hundreds more. For more information about Safari Books Online, please visit us online.

How to Contact Us

Please address comments and questions concerning this book to the publisher:

Make:
1160 Battery Street East, Suite 125
San Francisco, CA 94111
877-306-6253 (in the United States or Canada)
707-639-1355 (international or local)

Make: unites, inspires, informs, and entertains a growing community of resourceful people who undertake amazing projects in their backyards, basements, and garages. Make: celebrates your right to tweak, hack, and bend any technology to your will. The Make: audience continues to be a growing culture and community that believes in bettering ourselves, our environment, our educational system—our entire world. This is much more than an audience, it's a worldwide movement that Make: is leading—we call it the Maker Movement.

For more information about Make:, visit us online:

Make: magazine: *http://makezine.com/magazine*
Maker Faire: *http://makerfaire.com*
Makezine.com: *http://makezine.com*
Maker Shed: *http://makershed.com*

We have a web page for this book, where we list errata, examples, and any additional information. You can access this page at: *http://bit.ly/gsw-intel-edison*.

To comment or ask technical questions about this book, send email to *bookquestions@oreilly.com*.

Acknowledgements

There are a great many people I want to acknowledge for helping me, either directly or indirectly, with this book. First, the Intel Edison team, for supplying me with enumerable Edisons to play with over the past 12 months. Special thanks to Ed Ross, Jim

Chase, and Jay Melican for putting up with me during that time (and hopefully moving forward, too).

Second, to my bosses, Jeff Ota and Lakshman Krishnamurthy, who give me the time and freedom to play with Edison as a part of my actual job. To the remainder of the team, who challenge me every day to build something cool and do something with my life—thank you guys, too.

Super special thanks to my technical editors, Jason Wright (two thank yous!) and Esther Kim, for just being amazing people and friends and putting up with the many edits that this book needed. To Jonathan and Julija for playing the role of my non-technical editors and working through all the examples in the book (and also putting up with my needed edits).

Finally, thanks to my parents for letting me be a very curious child and indulging my need to take everything apart. And to my beautiful wife Kelsey, for tolerating my endless string of electronic and coding projects and my tendency to leave half-finished projects all over our otherwise clean house. I love you.

1/Introduction to Edison

The Intel Edison is an ultra-small computing platform that will change the way you look at embedded electronics. It's a powerful and adaptable piece of hardware that is compatible with a wide range of cutting-edge software solutions. Basically, Intel Edison is an entry-level Windows 8 tablet the size of a postage stamp that is sold for around $50 USD.

Intel Edison really shines for its small form and integrated wireless communications. For this reason, Edison is intended for embedded and connected devices. However, you'll see throughout this book that Edison's incredible computing power and flexible inputs and outputs enable a wide range of applications, from smart homes to self-driving robots to personal assistants.

So far we've been talking about the hardware inside the Intel Edison, but you'll be working with the integrated *Intel Edison compute module*. This module is shown in Figure 1-1 and can be accessed via a 70-pin connector on the bottom of the board (shown on the right side of Figure 1-1). While this connector is small and versatile, allowing developers to build custom boards that easily mate with the module, it is not meant for direct access. For this reason, a variety of *breakout boards* exist to help you get started. These boards break out the functionality of your Edison to a larger module that's easier to access.

Figure 1-1. *The front and back of the Intel Edison compute module*

Tour of Breakout Boards

If you're reading this book, there's a good chance that you've already purchased your Intel Edison compute module and an accompanying breakout board. While the compute module itself is standard, each breakout board is unique and will change the way you prototype and interface with Edison. For ease of use in this book, I'll be using the *Arduino Breakout Board*. This board is an Arduino-compatible breakout for Edison. I'll discuss what this means later in this chapter and a lot more in Chapter 3.

If you've not yet purchased this kit, then head over to one of the following retailers where you can buy your board. The cost is approximately $100 USD:

Maker Shed
    http://www.makershed.com/products/intel-edison-kit-for-arduino

SparkFun
    https://www.sparkfun.com/products/13097

Mouser Electronics
http://www.mouser.com/new/Intel/intel-edison/

Adafruit
http://www.adafruit.com/product/2180

Seeed
http://www.seeedstudio.com/depot/Intel-Edison-for-Arduino-p-2149.html

Amazon
http://amzn.com/B00ND1KH42

You might notice that the product pages all include an Intel Edison compute module with the Arduino breakout. Unfortunately, it is not possible to buy the breakout separately; the Arduino breakout always comes paired with an Edison. If you've already purchased a compute module or other breakout board, save it to later migrate your project to a smaller form factor. The Arduino board is the largest of the breakouts and is great for first-round prototypes.

The only other items necessary for getting started are two microUSB cables. Edison requires high-quality microUSB cables for powering the board; substandard cables (like the ones you often get for charging cell phones and other electronics) just won't cut it, especially if you're supplying power directly to the Edison via USB. MicroUSB cables can be found easily on Amazon or many other sites if you don't have them already. Spend a few extra dollars for the good ones.

At the beginning of subsequent chapters, I'll highlight any other hardware that's necessary to follow along. The full parts list for the entire book can be found in Appendix A.

The Arduino Breakout Board

The Arduino breakout with important labeled components is shown in Figure 1-2.

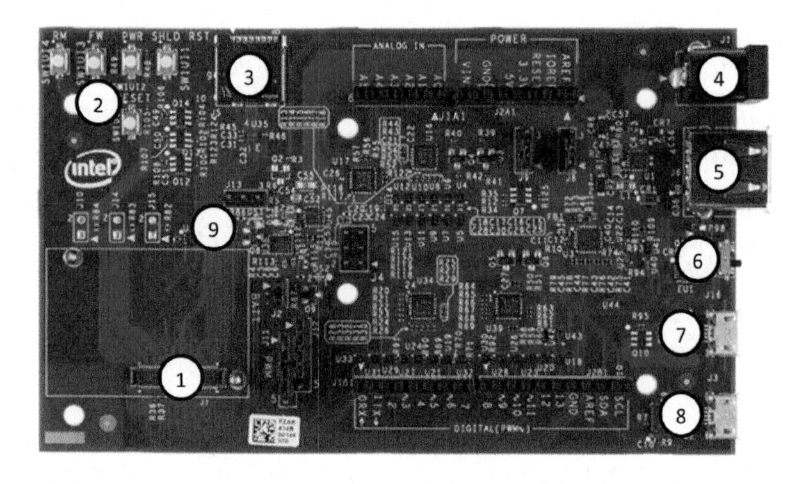

Figure 1-2. *Intel Edison Arduino Breakout Board with labeled components*

These important components are:

1. Female 70-pin connector. This connector mates with the male 70-pin connector on the back of the Edison compute module.
2. Buttons. These integrated buttons allow you to mechanically reset or power cycle your Intel Edison or the Arduino breakout.
3. MicroSD slot. While Intel Edison does come with 4 GB of integrated storage, the SD card slot gives the flexibility of adding additional storage to the system.
4. Barrel jack for power. This provides one option for powering your Edison. The barrel jack is the standard interface for a wall adapter power supply, which can be purchased for a few dollars on Amazon or SparkFun. Any supply between 7V and 15V with 500 mA or more will do. Alternatively, a battery with a barrel jack connector will also work, so long as it's in the same voltage range and can supply at least 500 mA continuously.
5. External USB port. Intel Edison has one full-size USB 2.0 port, selectable via the mechanical microswitch (labeled 6 in Figure 1-2). This port can be used to connect to peripheral

USB devices. If you wish to connect to multiple USB devices simultaneously, an externally powered USB hub is necessary.

6. Power mode microswitch. This switch allows you to toggle between using the full-size external USB port and the adjacent microUSB port (labeled 7 in Figure 1-2). The active port is the one that the switch is clicked toward. Note that you will need to supply external power if the USB port is selected, because the microUSB you'd otherwise use for power will be deactivated.
7. On-The-Go (OTG) microUSB port. This microUSB port supplies power and Ethernet capabilities, transfers Arduino sketches, and allows the host computer to access Edison as a mass-storage device.
8. Serial console USB port. Connecting this port to your host computer allows direct access to the Intel Edison operating system via serial communication. This port, however, does not supply power to the device.
9. Power LED indicator. When the device is powered on properly, this LED should remain in its lit state.

The Arduino breakout also has *input* and *output*, or I/O, pins that match the Arduino Uno in both physical layout and functionality. Input pins are used for signals that transmit information into your Intel Edison, such as the push of a button or measurements from a thermometer. Outputs are signals emitted by your Intel Edison, often to control attached devices. For instance, Edison might output voltage signals to make lights flicker or to display images on a screen. These I/O pins are labeled in Figure 1-3. Briefly, they are:

1. Digital input/output and pulse-width modulation
2. Analog input
3. Power
4. ICSP pins (for the Serial Peripheral Interface)

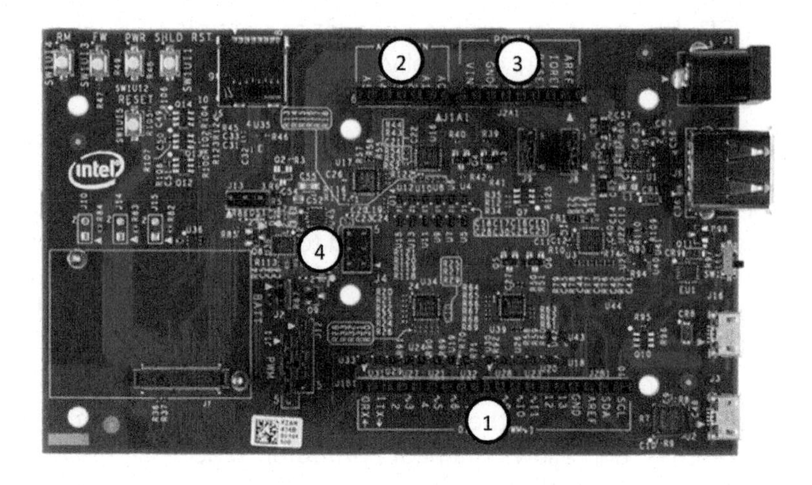

Figure 1-3. *Intel Edison Arduino Breakout with pins labeled*

Don't be worried if you're not familiar with these terms. You will be soon. In Chapter 3 and Chapter 4, you'll learn how to use and control these specific I/O pins. And, by using electronics that have been prebuilt for this configuration, I'll also illustrate why this layout is so extremely useful.

Other Breakout Options

Though you'll be using the Arduino Breakout Board in this book, it is worth taking a moment to explain the difference between this and the other breakout options: the *Intel Mini Breakout* Board and the *SparkFun Base Block*, shown in Figure 1-4 and Figure 1-5, respectively.

Figure 1-4. *Intel Mini Breakout Board and Intel Edison compute module*

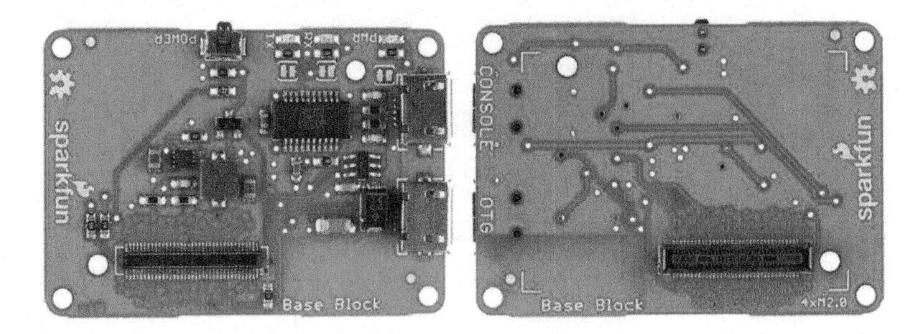

Figure 1-5. *SparkFun Base Block from the top and the bottom*

Both breakout boards provide access to the Edison compute module by attaching Edison to a female 70-pin connector. Both breakouts have two microUSB connectors: one for console access and one that serves as a USB On-the-Go (OTG) port. On the Intel mini breakout, the OTG port can also be used to supply power to the system. On the SparkFun Base Block, either micro-USB port can be used to supply power. Both boards have LED indicators lights for Edison power and charging.

There are two additional ways to supply power to the Intel mini breakout. The pair of header pins next to the microUSBs is for supplying 7-15V power. To more easily mate with a wall plug, a barrel jack can be soldered to the pads on the underside of this portion and used in lieu of the header pins. You can also use a 7-15V battery pack with a barrel jack connector if you want a battery-powered solution. Alternatively, the pair of pins on the opposite end of the board, labeled J22, can be connected to a 3.7V lithium polymer (LiPo) battery for power. A LiPo battery will supply enough power for most of the board functionality but will not provide enough to power peripherals attached to the micro-USB OTG port. The functional components of both boards are labeled in Figure 1-6 and Figure 1-7.

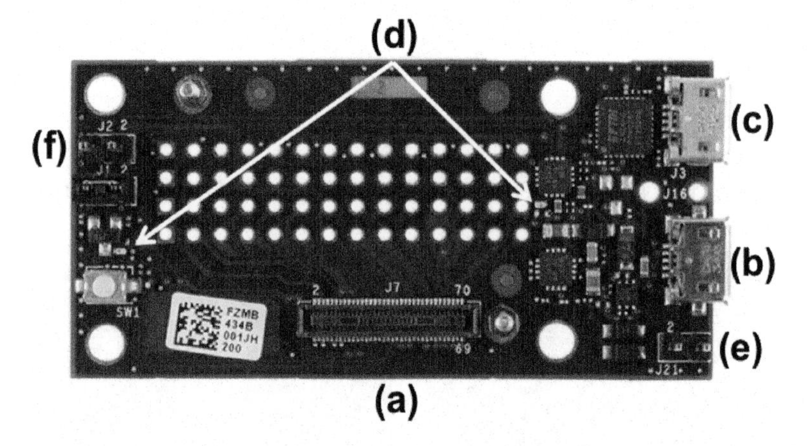

Figure 1-6. *Parts of the Intel Mini Breakout Board: (a) 70-pin connector, (b) OTG port, (c) serial console port, (d) power LEDs, (e) 7-15V input power headers, and (f) 3.7V LiPo battery headers*

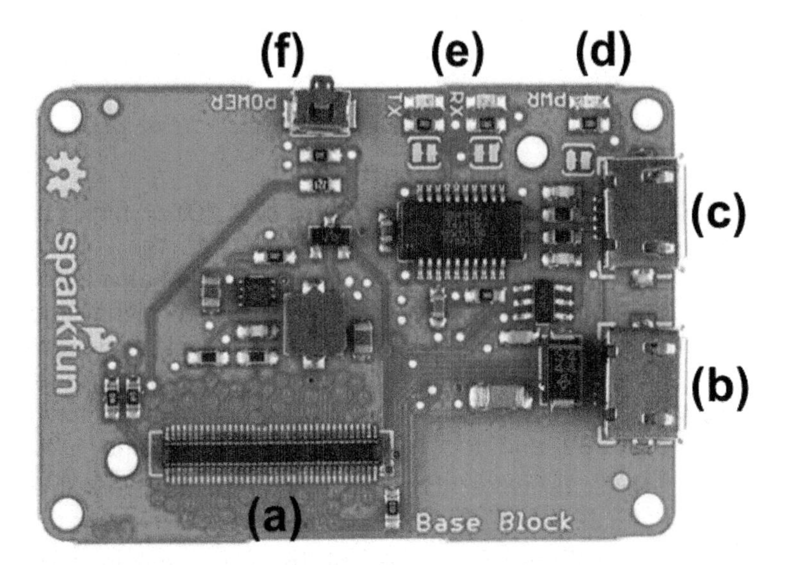

Figure 1-7. *Parts of the SparkFun Base Block: (a) 70-pin expansion header, (b) OTG port, (c) serial console port, (d) power LED, (e) data-transfer LEDs, and (f) power-reset button*

The main difference between the two breakouts is the way they use Edison's microcontroller. You'll notice that the Intel mini breakout has four rows of solderable pin holes, which are routed from the 70-pin connector. This allows you to access the programmable inputs and outputs on Edison directly. The SparkFun Base Block, on the other hand, provides no access to the I/Os. So, how do you attach sensors and other devices to the Edison?

If you look at the top of the Base Block (Figure 1-5), you'll notice that the 70-pin female connector runs through to a 70-pin male connector on the other side. SparkFun has made a series of modular boards for Edison, breaking out a few capabilities on each and designing them to stack using the 70-pin connectors all the way through. If you want access to a specific function, then you buy an additional breakout for that function and stack it with the base block. For example, if you want to read the values of a few digital inputs, you can buy the GPIO breakout. If you want to power your Edison using a LiPo battery, you can buy the battery block for stacking. And, of course, it is possible to stack more than two blocks together for increased functionality. The full list of SparkFun blocks can be found at SparkFun's website (*https://www.sparkfun.com/news/1589*).

Voltage Considerations

Digital signals have two states, low (0) or high (1), and the *voltage* of the signal refers to the voltage when that signal is high. Edison is a native 1.8V device, meaning it outputs a 1.8V high signal and expects to receive a 1.8V high signal back. A big reason why the Arduino breakout is so large is that it is stocked with *level shifters*, which convert between the 1.8V and the more common 3.3V and 5.0V (these voltages are typical of most sensors and peripheral devices in the maker space today).

Unlike the Arduino breakout, the mini breakout contains no level shifters and so won't be able to pass or read signals from higher voltage devices without you doing the level shifting off the board. Furthermore, if you happen to forget and send a higher voltage signal into the mini breakout, you can fry your Edison, since there's no on-board protection from higher voltages. This is the main reason why I use the Arduino breakout for the examples in this book. In Chapter 3 and Chapter 4, when you begin building systems with sensors and displays, you'll notice how much easier life is without worrying about shifting all the signals in and out. Be sure to check the voltage ratings for the different SparkFun breakout boards as you attach them, since each breakout board may have been designed for different voltage values.

Throughout the remainder of the book, wherever appropriate, I'll highlight when code will work as is or note the modifications necessary to port an example over to one of the smaller boards.

Setup and Configuration

Now that you've purchased your Edison and are comfortable staring at the Arduino breakout, let's power this guy up and get started! While Intel Edison is a full computer running a full operating system, it's meant for use in embedded products and does not have a graphical user interface. Everything you do for setup and programming will be done at the command line, making the startup and configuration process a bit more complicated than other maker products. Fear not. Once you get the hang of the command line, programming on Edison becomes a breeze. I take you on a whirlwind overview of command-line programming in Chapter 2, but for now, follow along to connect to and configure your Edison.

First, unpack everything from the box. The setup should be fairly intuitive; screw the four plastic posts into the corner, and click the Edison compute module into the 70-pin connector on the board by pressing down on the bottom of the module. It should snap in easily with a satisfying sound and feel. Secure the compute module using the two small screws provided with the kit. When you're finished, the assembled board should look like the one in Figure 1-8.

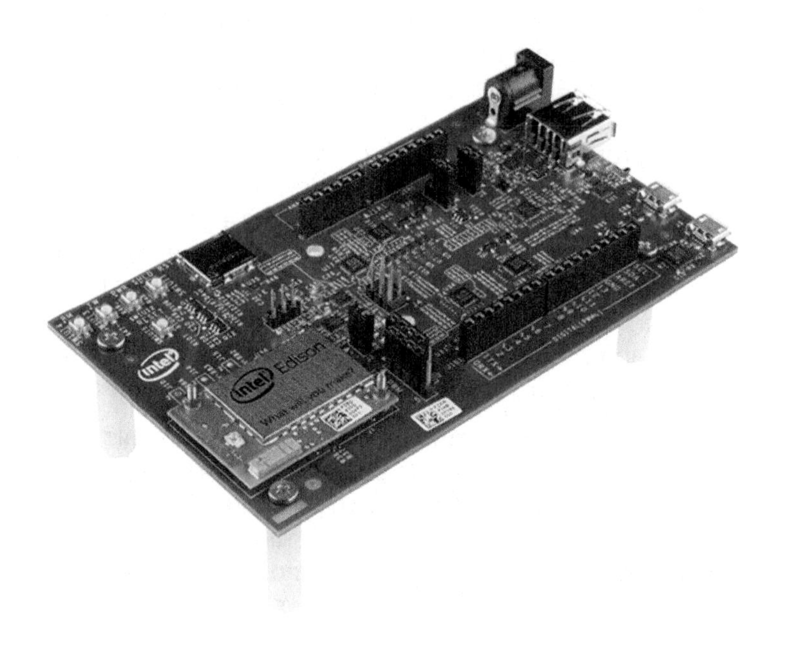

Figure 1-8. *Intel Edison assembled onto the Arduino Breakout Board*

The 70-pin connector is definitely a bit fragile, so be careful if and when you pop the compute module back off the Arduino breakout board. Try to avoid torquing it too hard, as the slotted pin connectors can bend.

Next, connect the Arduino breakout to your computer. First connect the Edison's power/external storage microUSB port (J16, closest to the power mode microswitch) to your computer. Once connected, the power LED indicator on the board will turn on, and Edison's mountable storage system should appear on your desktop as an external drive. If this does not happen, disconnect power, then make sure the switch near the USB and port is pushed toward the microUSB and that Edison is locked fully into the 70-pin connector. Try powering it up again.

Next, connect the other microUSB (J3) port to your computer. This will allow you to log in to the Edison operating system directly. A pictorial step-by-step walkthrough of this process can also be found in the Intel documentation (*http://bit.ly/intel-assemble*).

Install

Intel Edison's software requirements depend on your host computer operating system. Since the Edison release, Intel developers have worked very hard to make the installation process for Edison as seamless as possible. In the past few months, they've released a standalone installer for Windows and Mac that can be downloaded from their website (*https://software.intel.com/en-us/iot/library/edison-getting-started*). For Linux host computers, the process is a little more involved, but steps are laid out here and on the Intel website (*https://software.intel.com/en-us/get-started-edison-linux-step2*).

The processes described in this section will also flash Edison with its own most up-to-date operating system. Edison runs Yocto Linux by default, a configurable and lightweight embedded Linux OS. Because Intel developers are working hard to fix bugs and enable new Edison features all the time, the Yocto OS that comes on Edison straight out of the box is not the most up-to-date version.

Custom Yocto Builds

It is possible to build your own custom Yocto image and flash it onto your Edison. This is beyond the scope of this book, but there are several tutorials on this topic available online and a guide specifically for Intel Edison (*http://www.adafruit.com/datasheets/EdisonUG.pdf*).

Mac and Windows

Under "Step 2: Choose your Operating System," select your OS from the dropdown menu. and click the button that appears to start the download. Alternatively, you can download the stand-alone installers directly from the software downloads page (*https://software.intel.com/en-us/iot/hardware/edison/downloads*). The reason I include both is that Intel has a nasty little habit of moving its links and files around, and you'll sometimes find broken links to downloads. The software downloads page, however, tends to be very consistent and up to date.

The standalone installer will download and install the most recent Arduino IDE, the necessary drivers (if you're on Windows), and all other dependencies necessary for connecting to your Edison. Additionally, it will flash your Edison with the most up-to-date build.

Installation Options

When the installer asks you to configure the options, you're free to choose whether you want to install the last two options—Intel XDK IoT Edition and Eclipse. You won't be using these in this book, but they provide another fun way of interacting with and building programs for Intel Edison.

The first time the standalone installer is run, it will download quite a heavy load, so make sure you have a strong Internet connection before running it. If you do, the whole installation and flashing process should only take 10-15 minutes. Make sure to watch the installer as it runs, because it prompts you to take certain actions, such as plugging in the Arduino board, at different times during the install process. It also helps to be watching in case the installer throws an error, which has only happened to me during the board-flashing process.

If your flash fails, you'll either have to flash the board again with the standalone installer or perform a manual flash using the Phone Flash Tool. When run subsequent times, the installer will automatically detect the Edison-specific software on your system and skip the software download and install steps. However, in my experience with Edison, the Phone Flash Tool is much more reliable for flashing boards than the standalone installer, so it's definitely worth trying if you run into trouble. Instructions for performing this process can be found below the installer download button on the main page, as shown in Figure 1-9.

Figure 1-9. *Link location for performing a manual flash using Intel's Phone Flash Tool*

The Phone Flash Tool software, most recent Yocto complete image, and all the necessary drivers for install can also be found on the software page (*https://software.intel.com/en-us/iot/hardware/edison/downloads*).

Linux

Intel recommends using the Phone Flash Tool for flashing your Edison from a Linux host computer. The complete details for this process can be found in Steps 1-3 on the Intel setup tutorial (*https://software.intel.com/en-us/get-started-edison-linux*).

I've found, however, that it's easier to flash the board from a Linux computer using the old manual process. Basically, you'll use the Edison mounted storage device to load the newest build and then flash it directly on the board. Connect Edison to your host computer through both USB cables and wait until your host computer recognizes Edison as a mass-storage device. Then, open a new terminal on your host computer and remove everything that's on the storage device by issuing the following three commands at the command line. If you've never flashed the board before or used it to transfer files, you can skip this step:

```
cd /media/username/Edison
rm -rf *
rm -rf .*
```

Next, download the latest Yocto Complete Image from the Intel downloads page (*https://software.intel.com/en-us/iot/hardware/edison/downloads*). Unzip the files, and drag and drop them onto the Edison mounted storage device. After they finish copying, you're ready to flash.

To flash, you'll again use your terminal window. To connect to Edison, you'll use a program called screen. If you do not yet have screen installed on your computer, install it using your distribution's package manager. For example, under Debian and Ubuntu, you'd use the command:

```
sudo apt-get install screen
```

Now connect to Edison using screen by typing the following command:

```
sudo screen /dev/ttyUSB0 115200
```

Press Enter twice, and the login screen will be displayed. Log in using "root" as the username. At the Edison command prompt that displays, issue the command `reboot ota` and press Enter. Your Edison will reboot and begin flashing the newly loaded image. The flash should take a few minutes to complete. When it does, you will be prompted for your login again, which is still "root". Go ahead and log back in.

Connecting

The procedure for connecting to Edison depends on your computer operating system. Please follow the appropriate set of instructions that follows.

Windows

Open the Windows Device Manager, and expand the "Ports (COM and LPT)" dropdown as shown in Figure 1-10. Look for the

COM port associated with the "USB Serial Port." In Figure 1-10, this is COM6.

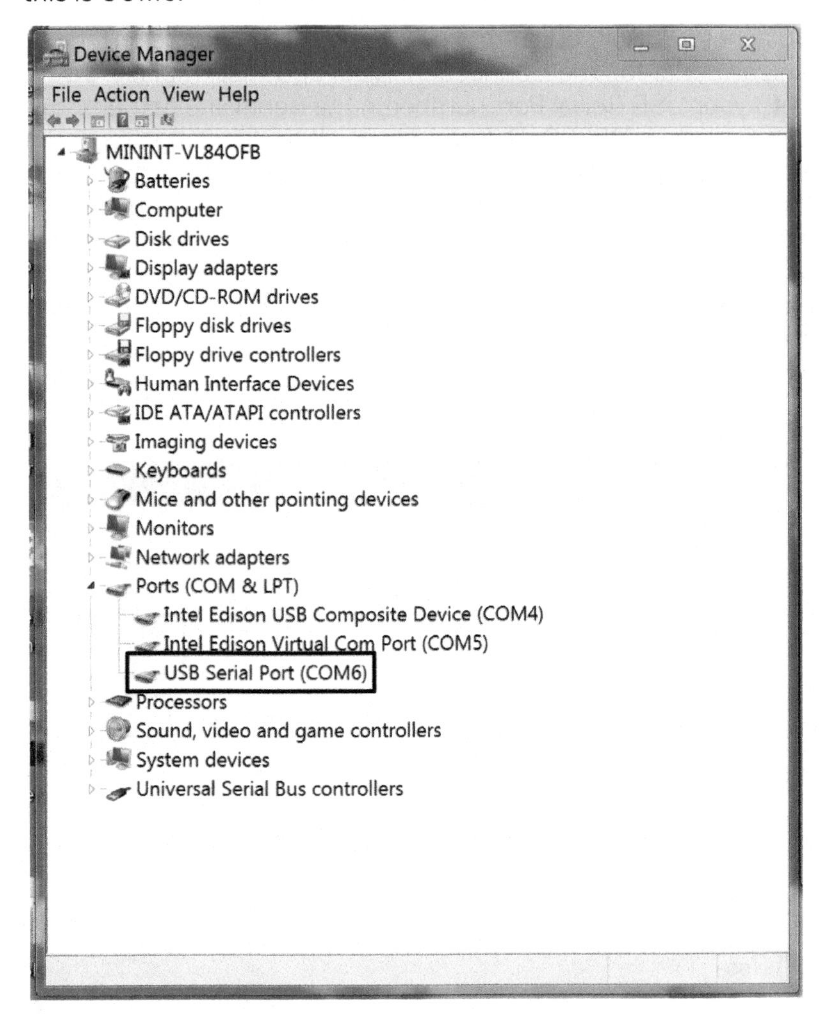

Figure 1-10. *Windows Device Manager for finding Edison serial communication port*

To connect to Edison interactively (to send commands at the boot prompt, for example) you'll need a terminal program. A good Windows option is PuTTY. You can download the PuTTY installer at the PuTTY download page (*http://bitly.com/putty-dl-*

page). Look for the file named *putty-0.64-installer.exe*, download it, and double-click it to install after it's finished downloading.

To access Edison, configure Putty as shown in Figure 1-11. First, select Serial for the Connection Type, and then replace COM6 with your USB Serial Port number in the Serial line text box. Set the Speed to 115200. Before opening the connection, you can save these parameters for future use by entering a name into the Saved Sessions text box and clicking the Save button as indicated in Figure 1-11. The name of your new saved session, which is "Edison" in Figure 1-11, should appear under "Default Settings." Click the Open button in the bottom right to initiate the connection to Edison.

To connect to Edison in the future, simply open Putty and double-click the name of your saved session.

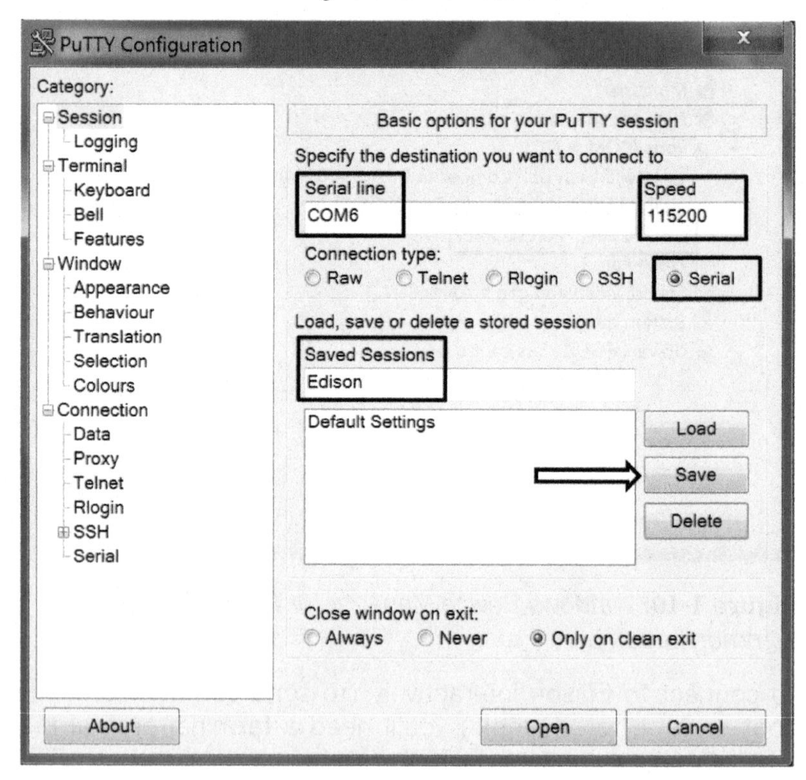

Figure 1-11. *Putty configuration for connecting to Intel Edison*

A blank terminal screen will appear after the connection is made. Press Enter twice, and you will be prompted for a login.

Mac

Open Terminal. If you've never used this program before, you can find it by searching for "terminal" in Mac Spotlight search. To connect to Edison, you'll use a program called screen. Type the following into the terminal and then press tab:

```
screen /dev/tty.usbserial-
```

You'll notice that the remainder of the Edison serial address autocompletes when you press tab. Complete the command by adding the configuration 115200 –L so that the entire line looks like this:

```
screen /dev/tty.usbserial-XXXXXXXX115200 -L
```

Press Enter twice, and you should see the Edison login prompt.

Linux

If you flashed your Edison using the old-fashioned method that I recommended, then you're already screen'ed into the device. If you used the Phone Flash Tool, then you still have to connect for a first time.

To connect, open a terminal window and install screen using your distribution's package manager. For example, under Debian and Ubuntu, you'd use the command:

```
sudo apt-get install screen
```

Now connect to Edison using screen with the following command:

```
sudo screen /dev/ttyUSB0 115200
```

Press Enter twice, and the login screen displays.

Navigating in Screen

If you've never used screen before, there are two important commands that you need to know: pressing Ctrl+A and then the K key will kill your current screen, and the terminal command `screen -r` will reattach you to a screen that is already in session.

Logging In

After connecting to Edison, you will be prompted for your login. The default login on Intel Edison is "root" with no associated password. Go ahead and enter the username now, and then press Enter. Your reward will be the Linux command line:

```
root@edison:~#
```

Congratulations! You're successfully connected to your Edison. If you're unsure of what to do at the Linux command line, don't worry. Chapter 2 provides a quick overview of commands to help you navigate and program your device.

Configuring Edison and Getting Online

At the Edison command prompt, type **`configure_edison --setup`**, and press Enter. Edison will walk you through the process of setting a password for the root user, naming your device, and connecting to the Internet. After connecting to the Internet, you will see the following response:

```
Done. Please connect your laptop or PC to the same network as
this device and go to http://XXX.XXX.X.XX or http://device
name.local in your browser.
```

Bad Builds

If you tried to issue the command `configure_edison --setup` and received an error message about `--setup` not being an option, then your build unfortunately failed. Take the actions recommended in the Install section for your host computer.

`XXX.XXX.X.XX` is your device IP address, and *`devicename`* is whatever you've just named your device. After configuration, Edison starts up its own node.js webserver, which is accessible from the browser on your host computer, so long as you're on the same network. Open a browser window and check it out. You'll see a webpage with the device name and IP address displayed against a blue background.

Firewalls and NATs

The automatic node.js server won't necessarily work if the Edison is connecting behind a firewall or network address translation service (NAT).

It is now possible to connect to Edison wirelessly instead of using screen, eliminating the need for a USB tether. Again, you must be on the same network in order for this to work.

To wirelessly connect from the terminal, use either of the following two commands:

```
ssh root@IP Address
ssh root@devicename.local
```

Then press Enter and provide your password when prompted.

If you're using Putty on Windows, Figure 1-12 shows the setup for a wireless connection.

To terminate the wireless connection on any system, either close the window or type **`exit`** at the Edison command prompt.

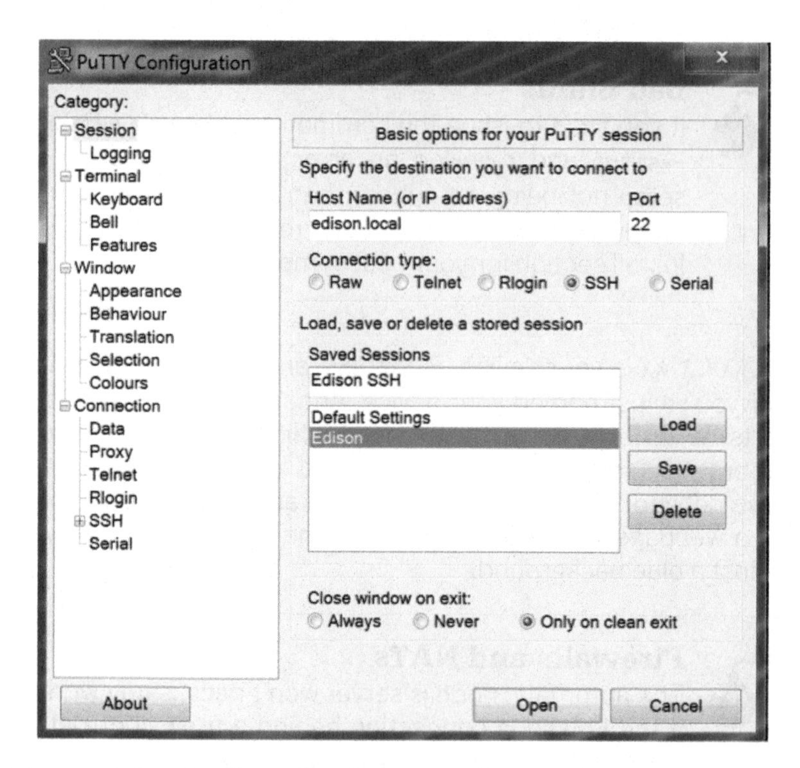

Figure 1-12. *Configuration for connecting wirelessly to Intel Edison using Putty; be sure to replace edison.local with your `device name`.local*

If you issue the command `configure_edison` or `configure_edison --help` at the Edison command prompt and hit Enter, you'll see the other configure options that are preinstalled in the current build. You can run any of these with the command `configure_edison` *`option`*. Some of the more important possibilities for *`option`* are:

`--wifi`
: Walks you through the process of configuring additional WiFi networks for Intel Edison. All networks added to Edison this way will be remembered and automatically joined at startup, with network preference going in the order in which they were added.

`--version`
: Returns the build version of the current Yocto system.

`--latest-version`
: Pings Intel's site for the newest officially released Yocto build for Edison.

`--upgrade`
: Downloads and installs the newest officially released Yocto build for Edison. Using this option is like performing an automatic flash of the most recent build. If you're already on the most recent build, this returns a status message and performs no action.

`--flash <version>`
: You can use this option to flash any previous or current version of the officially released Yocto builds. Replace <version> with any version number of interest.

Note that flashing Intel Edison will remove previous configuration settings, so you'll have to run **`configure_edison --setup`** to rename the device, set passwords, and configure WiFi each and every time you flash.

Finally, shutting down Intel Edison is easy: simply issue the following command and then press Enter:

```
shutdown now
```

Edison will display a series of shutdown messages ending with:

```
[  OK  ] Reached target Shutdown.
```

Then Edison will power itself off. The power LED on the Arduino board will turn off when the process is complete.

Troubleshooting

If you experience any trouble powering up the board, ensure that Edison is locked tightly into the 70-pin connector. If Edison still does not power on, try switching microUSB cables. A lot of cheaper cables can be current-limiting and might not be providing enough current to power Edison and the board. Finally, try switching to an external power supply using the barrel jack.

If you experience issues establishing tethered communication with the board, check that both microUSB cables are inserted fully. Make sure that you have selected the correct COM port or USB serial device address for your host computer and that the communication rate is set to 115200. You will still be able to connect with an incorrect rate, but you will receive only garbage printouts from Edison on your local machine.

If you experience issues with the wireless connection, make sure that Edison is still online, that your host computer is on the same network, and that you've correctly typed the IP address or device name when trying to connect.

If all else fails, the Intel Edison communities site (*https://communities.intel.com/community/makers/edison*) has heaps of questions and answers helping people to get set up with their boards.

Going Further

Intel Edison Downloads and Documentation (http://bit.ly/intel-dls)
: This is a great place to find more information about the hardware and software that make up the Intel Edison ecosystem.

Intel Edison Learning Center (http://bit.ly/intel-maker)
: A good, general starting point for learning to work with Intel Edison.

The Yocto Project (https://www.yoctoproject.org/)
: The definitive source for information about Yocto Linux and the Yocto Linux community.

The GNU Screen User's Manual (https://www.gnu.org/software/screen/manual/screen.html)
: An overview of useful and not-so-useful commands in the `screen` program. Aside from connecting to Intel Edison, screen is an extremely useful program for remote connection to other computers and devices.

2/Introduction to Linux

In June 2014, 97% of the world's fastest computers were powered by Linux. Linux is the operating system of choice for most of the servers powering the Internet and over a billion Android devices worldwide. And, of course, Linux is on your Intel Edison.

This chapter is meant to be a whirlwind tour of Linux and the specific Linux distribution running on Edison. By the end of the chapter, you'll be able to easily navigate your Edison filesystem, install packages, transfer files over the internet, and just feel comfortable in the driver's seat of the Linux command line.

What Is Linux?

Linux is a free and open source operating system (OS). It runs on top of the Linux kernel, which was first released by Linus Torvalds on October 5, 1991. It is fitting that Linux is the operating system running on Intel Edison, since the original intent of Linux was to serve as a free operating system for Intel x86-based personal computers.

Truly, Linux is one of the shining stars of the open source software movement. With over 10,000 contributing developers and over 17 million lines of source code, the estimated cost for the entire Linux code base is over $300 million USD. With so many different contributors and use cases, it's not surprising that many different flavors of Linux have emerged. You may already be familiar with some of them: Ubuntu, Arch, Fedora, and Gentoo. These different Linux distributions exist to suit different needs. Whether it's increased security, a faster/lighter operating system, or better usability, there's a flavor of Linux for every occasion.

The version of Linux running on Intel Edison is *Yocto Linux*. Yocto Linux is designed specifically for *embedded devices*, which are devices buried within the hardware of a system and often used to control the system in real time. Yocto is highly customizable; while all builds will contain the same core, you can build almost an infinite number of versions around the core to suit your hardware and software needs. For example, the Intel Edison version of Yocto is already built to enable Arduino-compatible hardware interfaces. And Intel is enabling new features and improving their build all the time, which is why we had to flash a new build in Chapter 1.

Because it's built for embedded devices, Yocto Linux does not have a traditional point-and-click user interface. In fact, there's no graphical user interface (GUI) at all, meaning we'll have to do 100% of our operations and programming on the command line if we're not using the Arduino IDE. Thus, to leverage Edison's full power, it becomes imperative that we know how to navigate the Linux command line properly. Let's start by explaining how the Linux filesystem works.

The Edison Filesystem

On your host computer, all files are stored in folders. In Linux, we call each of these folders a *directory*. And just like most of the folders on your host computer, directories on Linux often have a specific purpose and location.

Figure 2-1 shows a diagram with analagous directory structures between the Linux and Windows operating systems. Just as you can think of *C:/* as the highest directory in Windows, you can think of */* as the highest directory in Linux. And just like Windows, the *path*, or full, absolute location, of a directory in Linux builds on the layer before it.

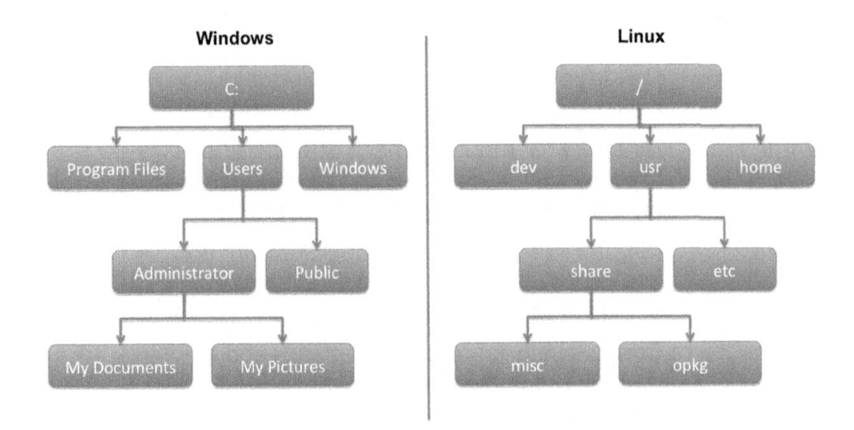

Figure 2-1. *A comparison of structure between the Windows and Linux file systems. Only a few directories and subdirectories are shown on each side for comparison.*

Some important paths on the Edison filesystem are shown in Table 2-1. Most of the paths match a standard Linux filesystem, though some, such as */sketch*, are unique to Edison.

Table 2-1. *Important paths and their functions on Intel Edison*

Directory path	Function
/bin	Contains binary executables comprising the basic system programs and utilities
/boot	Contains the files that the system needs to boot, including the Linux kernel, and system map.
/dev	Directory for physical and virtual devices on the system
/etc	Contains the Linux configurations files
/etc/init.d	Contains Linux startup scripts
/home/root	Home directory for the root user
/lib	Library files that can be accessed by the entire system
/lost+found	A place for Linux to store files that would normally be lost if there were damage to the filesystem, corruption, or loss.

Directory path	Function
/media	Preferred directory for attached storage devices, such as USB flash drives and SD cards
/mnt	Directory for mounting hard-drive partitions and physical devices, though */media* is often preferred for connected storage devices
/opt	Directory reserved for optional software and add-on packages that are not part of the default installation
/proc	Contains information and statistics about running processes and other runtime parameters
/sbin	Holds the binary executables comprising the basic system administrative programs and utilities
/sketch	Directory for holding compiled Arduino sketches
/sys	Contains information and statistics about Linux devices
/tmp	Contains temporary system files
/usr/bin	Contains additional system binaries
/var	Serves as a "catch-all" for data generated while the system is running
/var/log	Holds the Linux log files

Upon login to Edison, the command line is already greeting you with a set of information: your username (`root`), the device name (`edison`), and the path of your current working directory (~). Commands are typed to the right of the # symbol and executed by pressing Enter. An example of this command-line prompt is shown below:

```
root@edison~#
```

You may notice that ~ does not appear to be a match for any path we've mentioned so far. The symbol ~ represents your *home directory*, which the system has set to */home/root* by default. When you `ssh` or `screen` into Edison, you'll always start out there.

In future chapters, for brevity, the command line will be indicated simply by # or $ at the start of a command.

Intel Edison uses the Bourne shell, one of the many Linux shells out there. The Bourne shell supports tab-completion, meaning you only have to type the first few letters of a command or path and the shell will try to complete it for you when you press the Tab key. If multiple matches exist, the shell won't be able to complete but will show you the possible options if you hit Tab twice. Additionally, the shell keeps track of your command history, and you can access it by pressing the up arrow. This is extremely useful if you're repeatedly issuing the same set of a few commands.

If you're wirelessly connected to Edison, you can also exploit my two favorite command-line shortcuts: Ctrl+A and Ctrl+E. Ctrl+A will take you to the beginning of the current command-line prompt, right after the $. Ctrl+E will take you to the very end. These shortcuts are very useful if you're fixing a long command from your history that originally had typos. Instead of navigating the command with the left and right arrow keys, you can skip around with the Ctrl shortcuts. Note that this will not work if you're connected to Edison serially using the screen program. Ctrl+A is a screen command, and your host computer will register the keystrokes as such. Regardless of how you're connected to Edison, the all-powerful Ctrl+C will kill the command that is running in the terminal.

Basic Linux Commands

Let's now learn some commands to navigate the Edison filesystem. To change directories, use the cd (for "change directory") command. The following command will take you to the highest directory, /, on the filesystem:

```
root@edison~# cd /
```

To switch back to your home directory, you can use any of the following three commands:

```
root@edison/# cd /home/root
root@edison/# cd ~
root@edison/# cd
```

You can also change directories relative to your current location. Linux uses . to represent the current directory and .. to represent the folder one directory up in the filesystem. Issuing the following command will take you from the */home/root* directory to the */home* directory:

```
root@edison~# cd ..
root@edison:/home#
```

If you're ever wondering where on the filesystem you are, or if you want the full, absolute path to your current location, use the `pwd` (for "present working directory") command:

```
root@edison:~# pwd
/home/root
```

One of the most important commands in Linux is the `ls` (for "list directory contents") command, which lists the directory contents. To see the contents of the current directory, simply issue the command by itself:

```
root@edison~# ls
```

This command probably returned nothing, since we've not yet loaded any files or folders into our home directory. Let's look at a folder that will return some results. To list the contents of another directory, specify that path as shown in the following example:

```
root@edison:~# ls /etc/
bash_completion.d       issue.net               rc4.d
binfmt.d                ld.so.cache             rc5.d
and more...
```

The `ls` command can also be issued with *flags*. Flags set the configuration for any specific command and are issued after the name of the command itself. In Chapter 1, we used the flag `--setup` to tell the `configure_edison` command which options to configure. Most Linux commands support multiple flags, and the flag `--help` is often reserved to display the command help text. The `ls` flag `-a` tells Linux to show all files and folders in a directory, including hidden ones. Hidden files and directories begin with: `.`

```
root@edison:~# ls -a
.                .ash_history     .python-history
..               .node_app_slot
```

You can also set the flag `ls` to view the output in a list format. This provides additional information about permissions, file sizes, and dates.

```
root@edison:~# ls -al
drwxr-xr-x    4 root  root   4096 Mar 22 17:29 .
drwxr-xr-x    4 root  root   4096 Jan 21 17:14 ..
-rw-------    1 root  root    578 Mar 22 17:41 .ash_history
drwxr-xr-x    2 root  root   4096 Jan 21 17:14 .node_app_slot
-rw-------    1 root  root     13 Jan 21 17:23 .python-history
```

You can also add the `h` flag to increase "human readability," which shows the filesizes in a more appropriate format.

Flags

Some command-line utilities allow you to combine flags (like `-lah`) others don't. It takes a little playing with these commands to get used to this. Many flags also have a short-form version, a single - followed by single character, and long-form version, a -- followed by a word or words. For example, most commands have a help text that can be displayed with either the `-h` or the `--help` flag.

The command `touch` will create an empty file with the specified name:

```
root@edison:~# ls
root@edison:~# touch notes.txt
root@edison:~# ls
notes.txt
```

You can use the command `mkdir` to create a new, empty directory:

```
root@edison:~# mkdir newdirectory
root@edison:~# ls
newdirectory  notes.txt
```

You can copy files from one directory to the other with the copy command, cp. Specify the original file as the first command argument and then the copied file as the second. If you don't specify a filename in the second argument, the original file will be copied with the same name. If you specify a new filename, the file will be renamed when it is copied.

Tab Complete It!

When issuing the next few commands, remember to utilize your tab completion!

```
root@edison:~# ls newdirectory/
root@edison:~# cp notes.txt newdirectory/
root@edison:~# ls newdirectory/
notes.txt
root@edison:~# cp notes.txt newdirectory/newfilename.txt
root@edison:~# ls newdirectory/
newfilename.txt  notes.txt
```

To move a file from one directory to another, use the move command mv:

```
root@edison:~# mv notes.txt newdirectory/newfilename2.txt
root@edison:~# ls
newdirectory
root@edison:~# ls newdirectory/
newfilename.txt   newfilename2.txt  notes.txt
```

Notice that *notes.txt* is no longer present in the home directory. The mv command is like dragging and dropping a file into a folder on Mac or Windows, whereas cp is like copying the file and pasting it.

You can also use mv to rename a file in place:

```
root@edison:~# touch file1.txt
root@edison:~# ls
file1.txt     newdirectory
root@edison:~# mv file1.txt file2.txt
root@edison:~# ls
file2.txt     newdirectory
```

The Recursive Flag

One very important and almost universal flag is `-r`, which stands for recursive. Linux treats folders differently than it treats files and for this reason will not copy, move, or delete folders straight-away:

```
root@edison:~# cp newdirectory /home/
cp: omitting directory 'newdirectory'
```

The `-r` flag tells Linux to run the command not only for all files in the directory, but also for all folders, folders within folders, and files within folders recursively on until forever.

To delete a file, use the `rm` (for remove) command:

```
root@edison:~# ls
file2.txt     newdirectory
root@edison:~# rm file2.txt
root@edison:~# ls
newdirectory
```

Deleting Files

While Edison does have limited storage, making it necessary to delete files, use `rm` with caution and `rm -r` with extreme caution! Unlike your host computer, which moves deleted files to the trash can, the `rm` command is immediate and irreparable! Once those files are removed, they're gone forever.

I personally recommend creating a trash-can-like system for deletion on your Edison. Create a directory reserved for trash, and move the files that you're thinking of deleting into that directory instead of deleting them. Should your Edison run low on space, you can then delete the files in your trash can.

The command clear will clear the command prompt and give you a clean window. This is especially useful if you're focused on the output of a single command.

The command date will display your Edison's current date and time:

```
root@edison:~# date
Sun Mar 22 17:44:11 UTC 2015
```

The command df will give you a quick overview of your available disk space. If you start running out of space on your Edison, this command can help you decide how to reallocate. Use the flag -h, which stands for "human-readable," for an output that is easier to parse:

```
root@edison:~# df -h
Filesystem                Size      Used Available Use%
Mounted on
/dev/root                 1.4G    380.2M    961.1M  28% /
devtmpfs                479.9M         0    479.9M   0% /dev
```

There are two more basic commands worth mentioning. The first is find. Without a GUI search tool, it can often be difficult to find the file you're looking for. By itself, the find command will list the path to all files in the current working directory and in all subdirectories. Specifying a starting path and a search criteria can be extremely powerful. For example, the following command searches the entire filesystem (recursively from /) for a file named *libz.so*:

```
root@edison:~# find / -name libz.so
/usr/lib/libz.so
```

Additionally, you can search for text within files by using the grep command. The following example searches for the phrase "wpa" within the WiFi configuration file *wpa_supplicant.conf*:

```
root@edison:~# grep "wpa" /etc/wpa_supplicant/
wpa_supplicant.conf
ctrl_interface=/var/run/wpa_supplicant
```

For more flexible search criteria, you can use *regular expressions*. Regular expressions are an extremely powerful template matching system for words, numbers, and symbols. It is possible, using regular expressions, to limit searches and outputs by specifying the structure of the item you're looking for. While it is certainly beyond the scope of this book to explain the entire regular expression syntax (for more info, see the resources in "Going Further" on page 47), there's one specific key that's definitely worth mentioning, the wildcard *. At the Linux command line, * is your best friend. It allows you to match any number characters, and * by itself matches everything. The best way to illustrate how to use the * key is using examples:

Show all the files within the *newdirectory* folder:

```
root@edison:~# ls newdirectory/*
newdirectory/newfilename.txt   newdirectory/notes.txt
newdirectory/newfilename2.txt
```

List only those files starting with the letters *notes*:

```
root@edison:~# ls newdirectory/notes*
newdirectory/notes.txt
```

Find all files in the */var/run/* folder with the phrase "wpa" in the name:

```
root@edison:~# find /var/run/ -name *wpa*
/var/run/wpa_supplicant
```

Search all files in the */etc/wpa_supplicant/* folder for the phrase "each":

```
root@edison:~# grep "each" /etc/wpa_supplicant/*
/etc/wpa_supplicant/udhcpd-p2p.conf:# remaining for each lease
in the udhcpd leases file. This is
```

Accounts, Permissions, and Ownership

Every file on Edison has an owner and a specific set of permissions associated with it. Using `ls -l` will list permissions and other information, as shown in Figure 2-2.

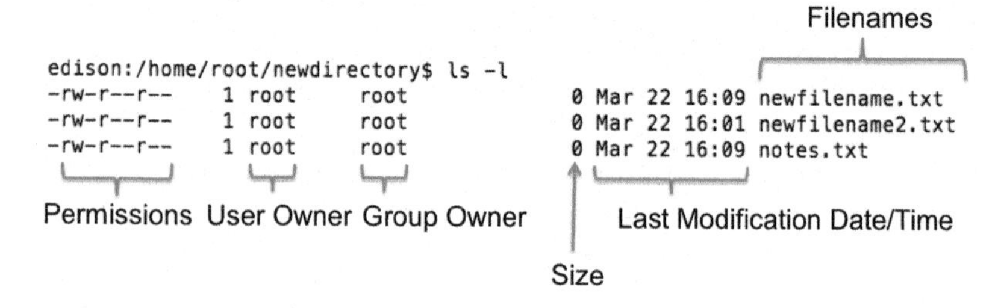

Figure 2-2. *Output of running* `ls -l`

Permissions are broken into three user based groups: owner, group, and all users. The owner is generally the person who has created the file or folder, unless someone else has taken control of it. Files and folders that came as part of the Linux operating system are owned by the root user by default. The group is the group of users that have been assigned to this file or directory. All-User permissions apply to every user of the filesystem.

As the root user, you can change and set ownership of a file by using the `chown` command:

```
root@edison:~/newdirectory# chown username filename
```

Each file or directory has three types of permissions: read, write, and execute. The read permission describes the user's ability to read the contents of the file. The write permission refers to a user's ability to modify or save a file or directory. The execute permission describes the user's ability to execute a file or view the contents of a directory.

Individual file and folder permissions are described by a dash followed by a set of nine letters: `-rwxrwxrwx`. The first three letters describe the owner's permissions to read, write, and execute the file or folder. The next three describe the group's read, write, and execute permissions; and the last three describe the same permissions for all users. Any letter that is replaced by a - is a permission that the corresponding user or group does *not* have. For example, in Figure 2-2, the `root` user is the owner of all three files and has the permission to both read and write, but not execute. The group and all other users have permission only to read these three files.

You can change these permissions using the `chmod` command. Specify the user or group with `u` for owner, `g` for group, and `a` or `o` for all users. Use the + sign to add permissions and the - sign to revoke them. Use `r`, `w`, and `x` to specify which permissions are being changed.

The following example gives write permission for the file *notes.txt* to all users:

```
root@edison:~/newdirectory# chmod a+w notes.txt
```

This example gives write and execute permissions for all files with the extension *.txt* to the groups that own them:

```
root@edison:~/newdirectory# chmod g+wx *.txt
```

And this example removes write and execute permissions for the owning group and read, write, and execute permissions for all users for any file starting with the word *new*:

```
root@edison:~/newdirectory# chmod g-wx,o-rwx new*
```

Currently, you're logged in as the root user, which is the superuser for your Intel Edison. Since default ownership of the Linux files goes to root, you have the supreme power to modify, move, delete, and own pretty much any file on the system. This also means that you can easily destroy necessary files and folders from the Edison system, rendering it useless. There are ways to recover from these mistakes, but it's best not to take chances. For this reason, it's often better to create a non-superuser account and use it for your prototyping and development.

To create a new user, use the command `useradd` followed by the new username:

```
root@edison:~# useradd username
```

You can change the password for this user with the `passwd` command followed by the username. Then, follow the prompts to set or change the password:

```
root@edison:~# passwd username
```

If you use `passwd` without specifying a username, Linux will prompt you for a new password for the logged-in account: in this case, root.

To log out of the account, press Ctrl+D. Go ahead and log in to the account you just created. You'll see a new command prompt, and you'll be started in your new home directory, */home/username*. Note that */home/root/* and subsequent files and folders will still exist on the system; it's just not the default home for your new user:

```
edison:~$ pwd
/home/username
edison:~$ cd ../root/newdirectory/
edison:/home/root/newdirectory$ ls
newfilename.txt   newfilename2.txt  notes.txt
```

To show the safegaurds built into the Linux system, we can try and delete a necessary file, such as our Internet configuration:

```
edison:~$ rm /etc/wpa_supplicant/wpa_supplicant.conf
rm: remove '/etc/wpa_supplicant/wpa_supplicant.conf'? y
rm: can't remove '/etc/wpa_supplicant/wpa_supplicant.conf':
Permission denied
```

Because we're not root, the system tells us that we don't have permission to perform such an action.

Scripting and More Advanced Linux Commands

Until now, we've been working with empty files created using the `touch` command. Let's fill them in with some content. The `echo` command simply repeats back any specified input and the `-e` flag tells the command to interpret line breaks and other similar characters:

```
root@edison:~# echo "1,2,3,4\n5,6,7,8"
1,2,3,4\n5,6,7,8
root@edison:~# echo -e "1,2,3,4\n5,6,7,8"
1,2,3,4
5,6,7,8
```

To redirect the command-line output to a file, use the `>` or `>>` symbols. If the specified file does not exist, it will be created. If the file already exists, using `>>` will append to the end of it, whereas `>` will replace the contents entirely:

```
root@edison:~# echo "1,2,3,4" > file1
root@edison:~# echo "5,6,7,8" >> file1
root@edison:~# ls -l > file2
```

To read the contents of this newly created file, use the `more` command:

```
root@edison:~# more file1
1,2,3,4
5,6,7,8
```

For longer files, you can press the Enter key to move one line down, the space bar to page down, and the `q` key to quit. Aside from `more`, you can also use `less`, which is similar but allows both forward and backward navigation through a file. Pressing `j` or Enter will move one line forward, pressing `k` will move one line backward, pressing the space bar will page down, and pressing `b` will page up. As with `more`, press `q` to quit.

Occasionally, you'll want to see only the first or last few lines of a file. This can be accomplished using the `head` and `tail` commands. By default, each of these commands will show you 10 lines, but this can be changed with the `-n` flag. For example, the following two commands show only the first and last lines of `file1`:

```
root@edison:~# head -n 1 file1
1,2,3,4
root@edison:~# tail -n 1 file1
5,6,7,8
```

The `cat` command also displays file contents but allows for multiple files to be specified:

```
root@edison:~# cat file1 file2
1,2,3,4
5,6,7,8
-rw-r--r--    1 root      root              16 Aug 26 16:43 file1
-rw-r--r--    1 root      root               0 Aug 26 16:44 file2
drwxr-xr-x    2 root      root            4096 Aug 26 16:39
newdirectory
```

You can also use `cat` to concatenate files together:

```
root@edison:~# cat file1 file2 > file3
root@edison:~# more file3
1,2,3,4
5,6,7,8
```

```
-rw-r--r--    1 root      root              16 Aug 26 16:43 file1
-rw-r--r--    1 root      root               0 Aug 26 16:44 file2
drwxr-xr-x    2 root      root            4096 Aug 26 16:39
newdirectory
```

Obviously, stringing these longer commands together gives you a lot of flexibility and power at the command line. Linux has another very elegant and simple way of chaining command-line commands together, known as *piping*. Piping takes the output of one command and supplies it as the input to another, using the | symbol as the pipe operator.

Let's use this with the `ps` command, which lists all running programs, their identification numbers (PIDs), and their stats. To find programs containing the word "launcher", pipe the output of `ps` into the `grep` command:

```
root@edison:~# ps | grep launcher
188 root       2404 S    {launcher.sh} /bin/sh /opt/edison/
launcher.sh
338 root       2412 S    grep launcher
```

The PID is the first number specified. To kill any running program on your system, use `kill` *`PID`*.

One fairly common debugging strategy you'll see on the Edison forums is to check out the end of the kernel message buffer by piping the `dmesg` command to `tail`:

```
root@edison:~# dmesg | tail
[  184.797665] CFGP2P-ERROR) wl_cfgp2p_add_p2p_disc_if : P2P
               interface registered
[  184.817868] WLC_E_IF: NO_IF set, event Ignored
[  232.514182] CFG80211-ERROR) wl_cfg80211_connect :
               Connectting with40:8b:07:67:f9:35 channel (11)
               ssid "Dante's Inferno", len (15)
[  232.514182]
[  232.600295] wl_bss_connect_done succeeded with
               40:8b:07:67:f9:35
[  232.612966] wl_bss_connect_done succeeded with
               40:8b:07:67:f9:35
[  235.092365] ip (530) used greatest stack depth: 4932 bytes
left
[  236.866492] systemd-fstab-generator[574]: Checking was
               requested for "rootfs", but it is not a device.
[  237.356150] systemd-fstab-generator[583]: Checking was
               requested for "rootfs", but it is not a device.
```

```
[  237.819098] systemd-fstab-generator[592]: Checking was
               requested for "rootfs", but it is not a device.
```

This is especially useful for checking if the last few actions you took actually worked. For example, if you just plugged in a USB device and it registered, you would see it in the last few lines of `dmesg`. Here, you can see that I just successfully connected to my home WiFi network, "Dante's Inferno."

Finally, you can write a series of command-line commands to a file with a `.sh` extension (required by the Bourne shell) and then execute all of them sequentially by running that file. This is extremely useful if you're performing a task that requires you to issue several commands repetitively:

```
root@edison:~# echo "ls -la" >> file3.sh
root@edison:~# echo "echo All Done" >> file3.sh
```

You can execute the script with the `sh` command:

```
root@edison:~# sh file3.sh
drwxr-xr-x    4 root      root    4096 Aug 26 16:59 .
drwxr-xr-x    4 root      root    4096 Jun 19 05:33 ..
drwxr-xr-x    2 root      root    4096 Jun 19
05:36 .node_app_slot
-rw-------    1 root      root      13 Jun 19 05:37 .python-
history
-rw-r--r--    1 root      root      16 Aug 26 16:43 file1
-rw-r--r--    1 root      root     196 Aug 26 16:44 file2
-rw-r--r--    1 root      root     212 Aug 26 16:51 file3
-rw-r--r--    1 root      root      21 Aug 26 16:59 file3.sh
drwxr-xr-x    2 root      root    4096 Aug 26 16:39 newdirectory
All Done
```

You can also change the permissions on the script to make it executable and run it directly:

```
root@edison:~# chmod +x file3.sh
root@edison:~# ./file3.sh
```

The Internet

If you configured your WiFi in "Configuring Edison and Getting Online" on page 20, then you should already be connected to the Internet. To check, you can use `ping` to send a packet of information to a site and acknowledge a response:

```
root@edison:~# ping google.com
```

You can transfer a URL or data to/from a server using `curl`. Curl is a very powerful tool that allows you to interact with the Internet from the command line—a necessary evil on Edison since there's no web browser. Issuing the following command will display the contents of the Google home page on the command line:

```
root@edison:~# curl http://www.google.com
```

Lynx

It's not strictly true that there are *no* web browsers available on a command-line-only system. Lynx (*https://kb.iu.edu/d/afik*) is a powerful and popular web browser for command-line interfaces. It does not come preinstalled on Edison, but if you want a command-line web browser, I recommend trying it.

Finally, a great way to pull files from the Internet is using the `wget` command:

```
root@edison:~# wget http://ftp.gnu.org/gnu/wget/
wget-1.5.3.tar.gz
```

`wget` can also be used to pull files recursively or from a list of sites.

Compressed Files

Many Linux software packages are distributed in a compressed format, often as `tar.gz` files. To view the files within, uncompress the package using the `tar` command and the `xvf` flag:

```
root@edison:~# tar xvf wget-1.5.3.tar.gz
```

To create a `tar.gz` file for upload or distribution, use the `cvf` flag, specify the name of the output file and then the files to be compressed:

```
root@edison:~# tar cvf data.tar.gz file3.sh get-
pip.py wget-1.5.3
```

You can also download files to your host computer and transfer them to Edison using `scp` or `sftp`. While you can do this from the command line, there are a number of great file transfer programs out there with nice graphical displays. My personal recommendation is CyberDuck (*https://cyberduck.io*) for Mac and Windows and FileZilla (*https://filezilla-project.org/download.php?show_all=1*) for Linux (though FileZilla actually works for all three platforms).

To initiate the connection in CyberDuck, click the Open Connection icon in the top-left corner. A new window will pop up asking for the connection parameters. Change the drop-down at the top from FTP (File Transfer Protocol to SFTP (SSH File Transfer Protocol). Then, in the server field, specify either Edison's IP address or ***devicename**.local* as well as the username and password for login in their respective fields. If you forgot your IP address, run the `configure_edison` command with the flag `--showWiFiIP`. This setup is shown in Figure 2-3. The setup process is very similar for FileZilla.

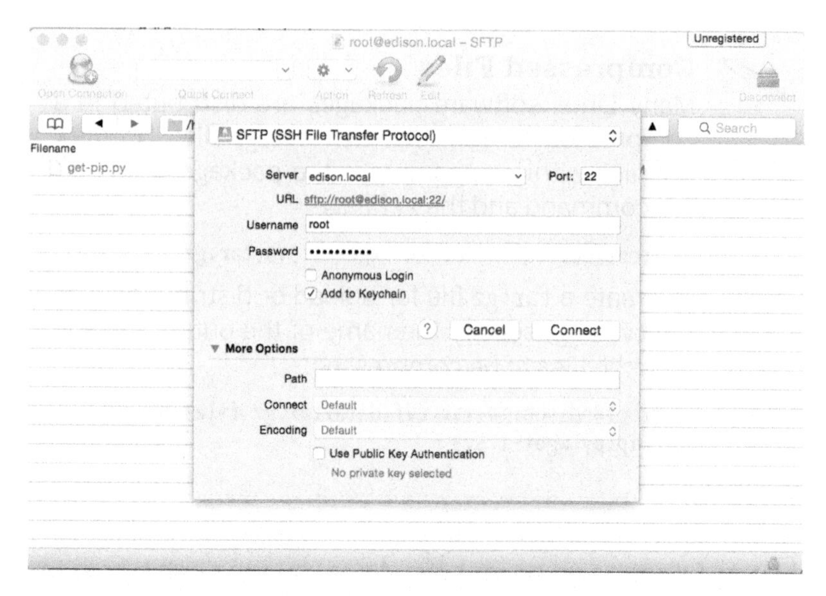

Figure 2-3. *Configuration for wireless file transfer using Putty; be sure to replace edison.local with your devicename.local or the device IP address*

Click the Connect button to make the connection. When the connection is established, the program starts you in the home directory by default. You can move up directories using the drop-down and double-click on folders to move through them. To transfer files in either direction, just drag and drop them between CyberDuck and your host computer.

Edison will only work with SFTP and not standard FTP, and SFTP will only work if you set a password for your Edison when you configured it. Otherwise, the connection will be blocked by the device.

Installing Packages in Yocto

Yocto ships with the `opkg` command-line tool for installing packages, which makes installing new packages quick and simple for a few reasons. First, `opkg` automatically installs all package *dependencies*, or software on which your new package relies. Second, the package installation is incredibly fast, as the files have already been precompiled and built for your Intel Edison.

Wherever possible, it's better to use the opkg utility for installing packages than downloading them from the Internet and installing them manually on Edison.

Before we can use opkg though, we have to edit the configuration files to tell it where to pull packages from. The opkg package stores its list of repositories in the */etc/opkg/base-feeds.conf* file. The following commands will append some useful Edison repositories to this file:

```
root@edison:~# echo -e "src/gz all http://repo.opkg.net/edison/
repo/all  \nsrc/gz edison http://repo.opkg.net/edison/repo/
edison \nsrc/gz core2-32 http://repo.opkg.net/edison/repo/
core2-32" >> /etc/opkg/base-feeds.conf
root@edison:~# echo "src mraa-upm http://iotdk.intel.com/
repos/1.1/intelgalactic" > /etc/opkg/mraa-upm.conf
```

Then, opkg will properly source these files when you tell it to update:

```
root@edison:~# opkg update
```

Going forward, any changes you make to these files will require an update to take effect.

After sourcing these files, do not perform an opkg upgrade, as this command will try to download all the available packages for install, quickly filling all available storage on your Edison.

Let's install a few packages that we'll need later: git, a web-based code-sharing and version control system, and mraa, Intel's I/O controller for Edison:

```
root@edison:~# opkg install git libmraa0
```

It's as easy as that!

To remove a package, use the remove command. The following is just an example removal; it is probably best to leave git on your system or reinstall it after the removal:

```
root@edison:~# opkg remove git
Removing package git from root...
```

If a package you're trying to remove is required by others on your system, opkg will issue a warning message and give you options for continuing. opkg absolutely will not let you remove packages that will crash other software without warning you first.

OPKG Packages

If you'd like to see all packages available to you, you can view the links we added to the configuration files. The most common software packages can be found here (*http://repo.opkg.net/edison/repo/core2-32*).

The package names look like *git_1.9.0-r0_core2-32.ipk*, but when you issue the `opkg install` command, only use the part of the package name up to the first _. So, for this package, it's simply `opkg install git`.

Text Editors

Up until now, you've been using echo and other command-line utilities to add text to files. Another option is to edit files directly using a text editor.

Without a graphical display, text editing can be a bit tough. The preinstalled text editor on Edison is *vi*. It's a powerful text-editor, but it definitely has a steep learning curve. Another GUI-less text editing option is nano, which is much easier for the new user. If you'd like to give it a try, install it now:

```
root@edison:~# opkg install nano
```

Whichever text editor you choose, use it now to turn off the logging functionality on Edison. While logging can be useful for debugging, the loggers on Edison run constantly and eat up a lot of usable storage space. To open any text file in your editor, issue the command:

```
root@edison:~# editor_name filename
```

The configuration file for logging is located at */etc/systemd/journald.conf*. Open the file, and change the line near the top from `Storage=persistent` to `Storage=none`.

Going Further

We've done so much, and we've still barely scratched the surface. If you're interested in diving deeper into any of these topics, you can use the following resources.

"Introduction to Linux" (http://tldp.org/LDP/intro-linux/html/)
: A great introductory guide to pretty much all aspects of Linux.

The Yocto Project (https://www.yoctoproject.org/)
: The maintainers of Yocto Linux; this site has links and discussions on pretty much everything Yocto-related.

"The Command-Line Cheat Sheet" (http://bit.ly/linux-cheat)
: A pretty comprehensive list of all the commands you might need on the Linux shell. Also touches on building software and monitoring processes.

Bourne Shell Programming (http://www.ooblick.com/text/sh/)
: In-depth scripting, including variables and functions, on the Bourne Shell.

Regular-Expressions.info (http://www.regular-expressions.info/)
: A list of tutorials for different regular expressions topics.

"An Extremely Quick and Simple Introduction to the Vi Text Editor" (http://bit.ly/vi-intro)
: An incredibly good overview introduction to the text editor vi.

:nano Command Manual (http://www.nano-editor.org/dist/v2.2/nano.html)
: The complete list of nano commands.

3/Introduction to Arduino

The maker movement, coined the "New Industrial Revolution," is taking the world by storm. Creating new electronics is no longer limited to high-tech laboratories and corporate funding; it's been brought directly into people's lives and homes via cheap electronic components, testbeds, and widely supported community spaces. It's high-school shop class meets the electronics industry at a million miles per hour, and a lot of it is the result of one simple idea: the Arduino.

What is Arduino

Arduino is a paired hardware and software platform that enables rapid design, prototyping, and building of devices that can sense and control aspects of the physical world. The hardware behind Arduino is an open source physical computing platform based on a simple Atmel microcontroller. A microcontroller is a small computer with its own core, memory, and periperals, including inputs and ouputs (I/Os), and USB connections. The software that controls the Arduino hardware is written to the board through an integrated development environment (IDE). The IDE makes it easy to program Arduino not only because it connects to the board, but also because it comes with a wealth of examples and preprogrammed functionality.

Materials List

This chapter, along with Chapter 4, requires the most parts in the book. They are listed below as well as in Appendix A. An alternative to buying the first six items in this list separately is to purchase an Arduino-style starter pack, such as the SunFounder Sidekick Basic Starter Kit (*http://bit.ly/sidekick-kit*). Just make sure that the one you purchase contains all of the first six items (down to potentiometer):

A breadboard
: Although any solderless breadboard will do, it's nice to buy one with power rails on the sides (see "The Blink Circuit" on page 61 for further explanation). For all the exercises in this chapter, a half-size breadboard will give you more than enough area to work with. These can be purchased for a few dollars on Amazon (*http://bit.ly/BB400-board*). You can also purchase them from Adafruit (*http://www.adafruit.com/product/64*), but they're slightly more expensive at $5.

Male-to-male breadboard wires
: Used for connecting Edison's pins to your breadboard; make sure you buy male to male (M/M). These are probably cheapest on Amazon (*http://bit.ly/j-wires-mm*). Adafruit also sells packs in two different flavors: regular in mixed lengths (*http://www.adafruit.com/product/153*) and premium in just one length (*https://www.adafruit.com/products/760*). The range in price is $6-$8.

An LED
: To complete the exercises in this chapter, you really only need a single bare LED, but it's almost impossible to buy an LED as an individual unit. That's OK! LEDs are super useful for all manner of experiments. Adafruit has heaps of them (*http://www.adafruit.com/categories/90*), and you're looking specifically for ones like this (*http://www.adafruit.com/products/299*), a bare breadboard LED in a single color. You can also buy mixed packs on Amazon (*http://bit.ly/2pin-led*).

Resistors
: A pack of resistors (*http://www.amazon.com/dp/B0111SOTP6?psc=1*) with any value from 220 ohms to 1 kohms will work for this book. However, if you're thinking about getting more seriously into building and making, try getting a mixed pack in a large range, such as the E-Projects - 400 Piece, 16 Value Resistor Kit (*http://bit.ly/16v-resistor*).

A button
: The cheapest and easiest are sold as a pack on Adafruit (*https://www.adafruit.com/products/367*) for under $3. However, button choices are fairly limitless, and you can buy some really cool ones (*http://www.adafruit.com/category/235*) if you're so inclined.

A potentiometer
: You can buy a single potentiometer from Adafruit (*https://www.adafruit.com/products/356*) for under $2.

An accelerometer
: The specific accelerometer chip you'll be using in this chapter is the MMA8451 sold for under $10 on Adafruit (*http://www.adafruit.com/products/2019*). Accelerometers are one of the most widely used and basic sensors for embedded devices, which is why I chose it to pair with Edison.

A SPI-driven display
: Any screen that uses ILI9341 will work with the exercises in this book. You can find some cheap ones on eBay and Newegg, but I personally recommend the resistive touch shield (*http://www.adafruit.com/products/1651*) (more on why in "SPI Screen" on page 87) or the resistive touch breakout (*http://www.adafruit.com/products/1770*).

You'll need a soldering iron to solder the pin headers onto the accelerometer and the latter of the two SPI screens listed above.

The Arduino IDE

The Arduino IDE is the conduit to programming your Arduino board. It makes it easy to write code and upload it to the board, and runs on Windows, Mac OS X, and Linux. Like everything Arduino, the IDE is open source.

Installing

Intel Edison is an Arduino-compatible device, meaning it's not an Arduino but can be used almost identically as one. Edison uses an Intel Quark microcontroller to drive the I/Os on the Edison board. When originally released, Edison required a separate IDE than Arduino boards. Since then, both the Intel and Arduino folks have gone to great lengths to integrate the functionality of both into one IDE. From version 1.6 onward, the IDE distributed by Intel and Arduino will work with both platforms.

If your host computer is a Linux machine or you skipped straight to flashing your Edison with the Phone Flash Tool in Chapter 1, then you need to install the Arduino IDE before continuing. To download and install this IDE, go to the Intel (*http://www.intel.com/support/edison/sb/CS-035180.htm*) or Arduino (*http://www.arduino.cc/en/main/Software*) download page and click the link for your host computer system. After the download completes, unpacking the file creates the IDE for your platform. If you're on Mac or Linux, move it to a convenient location on your host computer. For instance, on a Mac, you might want to place it in the Applications folder. Double-click the IDE to start it.

Otherwise, the IDE is already installed on your computer. Locate it and open the program.

Navigating the IDE

The Arduino IDE is shown in Figure 3-1. The IDE is broken into several regions, the most important of which is the large white space that contains your code, also known as a *sketch*.

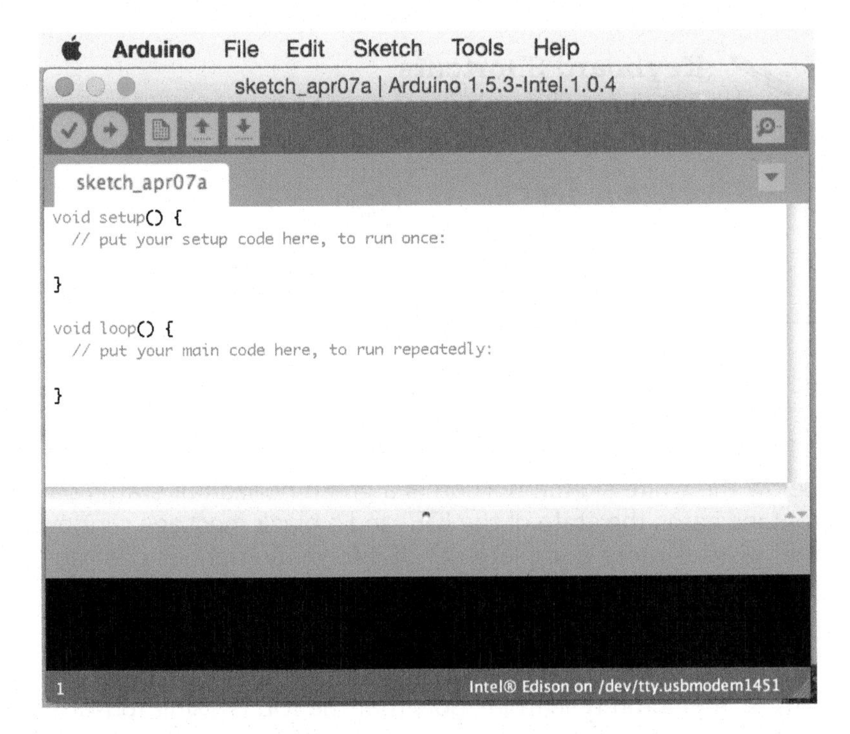

Figure 3-1. *The Arduino IDE for Intel Edison*

To the top-left of the sketch window are five light-green buttons. From left to right, they are:

Verify
: Saves your sketch and checks it for errors

Upload
: Verifies your code, transfers it to Intel Edison, and runs it on the board

New
: Creates a new IDE window with a blank sketch

Open
: Opens an existing sketch

Save
: Saves the current sketch

Keyboard Shortcuts

The preceeding commands also have some useful keyboard shortcuts:

- Ctrl+U or Cmd+U: Upload
- Ctrl+R or Cmd+R: Verify
- Ctrl+S or Cmd+S: Save
- Ctrl+N or Cmd+N: New
- Ctrl+O or Cmd+O: Open

Below the white sketch portion is a green horizontal status bar that indicates the state of the IDE, and a black message window that gives a more complete set of information about past and present status. Both of these portions of the IDE should be blank. If you click the Upload button at the top, you'll notice that they become populated with information.

Finally, the communications port that the IDE is using to transmit information to and from your Intel Edison is shown in the bottom-right corner.

Before you begin programming, you need to configure your IDE to communicate properly with our Edison. This means selecting the proper board and communications port.

To select your Edison, open the Tools drop-down, select Board, and then Intel Edison. To select your communications port, open the Tools drop-down, select Port, and then the port of the form /dev/cu.usbmodem*XXXX* for Mac or COM*XX* for Windows. One of the easiest ways to find which port belongs to your Edison is check the available ports with Edison unplugged and then plug it in and choose the new port that wasn't there before.

If Edison does not appear in your board selection menu, then you'll need to add it via the Boards Manager. Go to Tools → Board → Board Manager... Click on the Intel i686 Boards box and then click Install. When complete, Edison will appear in the "Board" drop-down.

Sketches and Functions

Every Arduino sketch requires two *functions*: *setup* and *loop*. A function is a self-contained piece of code that runs independently of the rest of the program. The setup function runs exactly one time when the program starts. After the setup function completes, the loop function runs over and over again so long as Edison is powered. Think about it like going for a run. The setup function is like putting on your running shoes: you do this exactly one time at the beginning of your run. The loop function is like the run itself: you take step after step after step until your run completes and you "power down."

Arduino is based on the programming language C++. In C++, you define your functions with the same structure every time:

```
return_type function_name (input variables here) {
    //Code here
}
```

The first word of your function states the type of output or return from that function. Because the setup function does not return any output, it has the return type **void**. The second word is the name of the function and can be made up of any combination of alphanumeric characters and the underscore (_), so long as the first character is not a number. The input variables follow the function name. They are surrounded by parenthesis and separated by commas. For the function named setup, the parentheses are left empty since the function does not take any inputs. Finally, all code associated with a given function is placed between curly brackets. The opening curly bracket comes after the closing parenthesis around the variables, and the closing curly bracket comes after all the code is complete.

Comments

Comments are like annotation for code. They themselves are not code; they are written into code to help increase understanding and readability. There are two ways to write comments in C++:

// Single line
: Everything on this line is now a comment

/ Multi-line*
: Everything is a comment until the closing symbols */

You see a single line comment (`//Code here`) in the previous code samples.

There are many native types in C++ and they're used for declarations as well as function definition. Some of the more common types we'll use are:

boolean
: binary true or false

byte
: an integer from 0 to 255

char
: an integer from -128 to 127

unsigned char
: same values as byte

word
: integer value from 0 to 65535

unsigned int
: same values as word

int
: integer value from -32768 to 32767. This is the type that is most commonly seen in Arduino examples and the type you'll probably use the most.

unsigned long
integer value from 0 to 4,294,967,295

long
integer value from -2,147,483,648 to 2,147,483,647

float
decimal value from -3.4028235E38 to 3.4028235E38

Digital Output with Blink

It's easier to understand all of this through an example. Let's start with the quintessential first microcontroller program, blink. In this example, you'll use Edison to control the output of a *light emitting diode* (LED), a semiconductor device that emits light when a current passes through it. This is the simplest circuit you can control with your Edison to observe a physical output.

The Arduino IDE provides example code, and we'll use this to get started. Open the File drop-down, then select Examples, 01.Basics, Blink. In the new window that opens, you'll notice that the `setup` and `loop` functions are now populated with code:

```
// the setup function runs once when you press reset or power
// the board
void setup() {     ❶
  // initialize digital pin 13 as an output.
  pinMode(13, OUTPUT);
}

// the loop function runs over and over again forever
void loop() {     ❷
  digitalWrite(13, HIGH);    // turn the LED on (HIGH is
                             // the voltage level)
  delay(1000);               // wait for a second
  digitalWrite(13, LOW);     // turn the LED off by making
                             // the voltage LOW
  delay(1000);               // wait for a second
}
```

❶ Within the `setup` function, we call the function `pinMode`, which takes two arguments. The first argument is an integer specifying the pin number. The pin numbers are written next to each of the Arduino-compatible pins on the top of the board. The second argument tells the hardware in which mode to operate the pin: input or output. Since an LED is an output device (it outputs light), we set the pin attached to the LED to reflect this. Pin 13 is special; while it does have a slot for connecting external hardware, it's also wired to an LED directly on the board. This is the LED that your code will make blink.

❷ The `loop` function issues four sequential commands that toggle the LED on and off at one-second intervals. Pin 13 is a digital pin, meaning it only takes one of two possible values: `HIGH` or `LOW`. You use the `digitalWrite` function to control the output of digital pins. When you write pin 13 to `HIGH`, you're telling Edison to send a high voltage (5 volts on the Arduino breakout) signal to pin 13, lighting the LED. Conversely, when you write pin 13 to LOW, you're telling Edison to send a zero voltage signal to pin 13, turning the LED off. In between, we use the delay function to pause the program. The argument `1000` tells delay to wait for 1,000 milliseconds, or 1 second. You can generate even shorter delays using the `delayMicroseconds` function.

Don't Forget the SemiColon

All commands and variable declarations in C++ are terminated with a semicolon.

Upload this code. You'll notice that an LED on the board begins to blink on and off every second. This is the on-board LED that is wired to pin 13, and you're controlling it with your Intel Edison!

One of the aspects that makes Arduino so convenient is that functions like `digitalWrite`, `pinMode`, and `delay` are so common that they've been preprogrammed into the Arduino software environment so that you don't have to write them yourself.

Saving Sketches

The Arduino IDE creates an Arduino folder within the Documents folder on Mac, and Arduino folder within My Documents on Windows, and a Sketchbook folder */home/username/* on Linux. By default, any sketches that you save will be saved to this directory. If you want, you can change this by editing your preferences in the Arduino IDE.

Going Further with Blink

Now you'll expand your blink example a bit. First, you'll define some *variables* to make your code read a little easier. A variable is a reference to stored information and the name is typically chosen to represent the information it contains. Declare the variable called `ledPin` at the top of the sketch, before the setup function. Since the ledPin number is an integer, we'll declare it as such:

```
int ledPin = 13;
```

Declaring the variable in the main part of the sketch, outside of any functions, means that every piece of code within your sketch has access to it. Change the number 13 in `pinMode` and `digitalWrite` to our variable `ledPin`. You can also declare a variable for the `delayTime` and specify this in the `delay` function.

```
int delayTime = 1000;
```

Aside from variables making it easier to simply read what's going on, they also make your code much more easily adaptable. Suppose you wanted to blink the LED at a cadence of every five seconds. Now all you need to do is change the variable at the top instead of adjusting the numbers all throughout your code.

When you specify `HIGH` and `LOW` to **`digitalWrite`**, you're actually using variables that have been predefined in the Arduino software. `HIGH` is an integer variable equal to one, and `LOW` is an integer variable equal to zero.

Finally, you're going to add a function that handles the blinking. Name this function **`blinkIt`**, and since it has no return values, declare this function with **`void`**. Move the contents of **`loop`** into the function **`blinkIt`**:

```
void blinkIt() {
  digitalWrite(ledPin, HIGH);   // turn the LED on (HIGH
                                // is the voltage level)
  delay(delayTime);             // wait for a second
  digitalWrite(ledPin, LOW);    // turn the LED off by making
                                // the voltage LOW
  delay(delayTime);             // wait for a second
}
```

Since the **`blinkIt`** function will perform all the steps it takes to make your Edison board blink, all you need to do now is call it in the **`loop`** function to run it over and over again forever:

```
void loop() {
  blinkIt();
}
```

Your final code should look something like this:

```
int ledPin = 13;
int delayTime = 1000;

void blinkIt() {
  digitalWrite(ledPin, HIGH);
  delay(delayTime);
  digitalWrite(ledPin, LOW);
  delay(delayTime);
}

void setup() {
  pinMode(ledPin, OUTPUT);
}
```

```
void loop() {
  blinkIt();
}
```

Upload this code to your Edison, and you'll see that your board keeps on blinking. Try playing with the delay times to see what Edison and your eyes can handle.

The Blink Circuit

Blink is not only a great introduction to Arduino programming, but also an excellent introduction to circuits. You're now going to move your blink circuit off the board and onto a breadboard.

Circuits consist of closed loops of electrical current. The current originates at a *source*, such as a battery or power supply, and terminates at the *return* or *ground*. The pins on your Arduino breakout can act as both sources and returns. When the circuit loop is complete, the electrons that make up the current are returned back to the source. Any part of an electrical circuit between the source and return is called the *load*. The load in an electrical circuit can be pretty much anything: lights, resistors, screens, speakers, and even the wire making the connections.

Some common circuit elements, including all the ones that you'll use in this book, are shown in Figure 3-2. The top row of Figure 3-2 shows how these common parts are often represented in drawings. The bottom row shows how these parts are commonly represented in a *circuit diagram*, or graphical representation of an electrical circuit.

From left to right, these elements are (a) resistors, (b) a potentiometer, (c) LEDs, and (d) pushbuttons. Ideal resistors are electronic components that have a specific, never-changing resistance to the movement of current through them. Resistors come in many different sizes, shapes, and values. You'll notice that there are three resistors in Figure 3-2 (a). These resistors have different resistance values (220 ohms, 470 ohms, and 1 kohm) that can be read from their color codes. We'll talk in more depth about resistors and the rest of these items later on in this chapter. For the representations of many more common components in circuit diagrams, see SparkFun's tutorial (*https://learn.sparkfun.com/tutorials/how-to-read-a-schematic*).

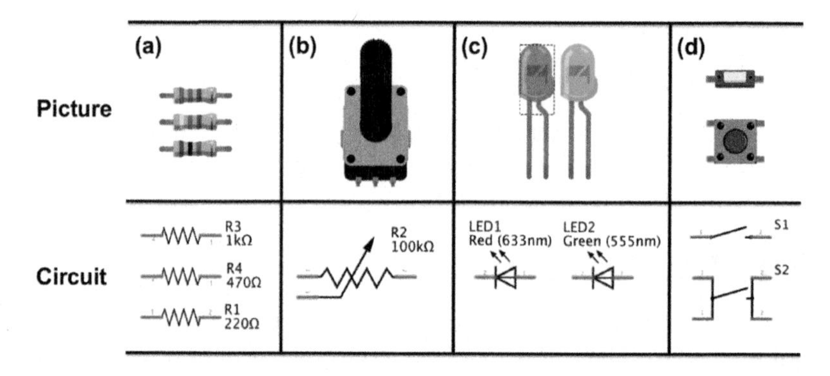

Figure 3-2. *Table of common circuit elements you'll use in this book. The top row shows the pictorial representation of each part, and the bottom row shows possible circuit diagram representations.*

Power Down First

Basic electrical safety requires removing power from the circuit before changing connections. This will avoid damaging your board, your components, and (most importantly) yourself.

For your blink circuit, you'll need some jumper wires, a single color LED, a resistor in the range of 220 ohm to 1 kohms, and a solder-less breadboard.

The inside of a solder-less breadboard is shown in Figure 3-3. Pockets of metal run across the rows, electrically connecting everything you push into the same row automatically. Rows on opposite sides of the center divider are kept separate. Your breadboard might also have two differently spaced columns along each side, often surrounded by red and blue lines. These are known as *rails*. On the rails, the electronic connection is throughout the whole column instead of across the row. The electronic signal across the two rails is typically reserved for power (red) and ground (blue/black). Electrically connected rows in a solderless breadboard are shown in Figure 3-3.

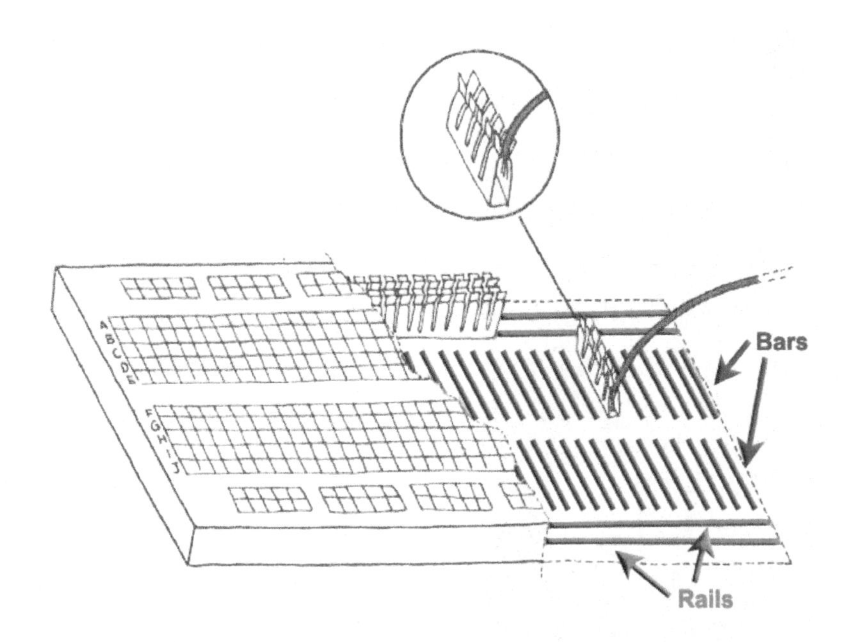

Figure 3-3. *The electrical connections on the inside of a solderless breadboard*

Wire your circuit as shown in Figure 3-4. Notice that the breadboard makes several connections between components for you and that your Intel Edison completes the closed loop of your circuit, acting both as source (output of pin 12) and return (ground). LEDs are directional devices, meaning they have an *anode*, or higher voltage side, and a *cathode*, or lower voltage side. Typical new LEDs have the anode on the longer leg, so make sure that the long leg of your LED is toward pin 12 (the source and therefore higher voltage side) and the short leg is closer to ground (the sink, therefore lower voltage side), or your LED won't light up when you start the circuit.

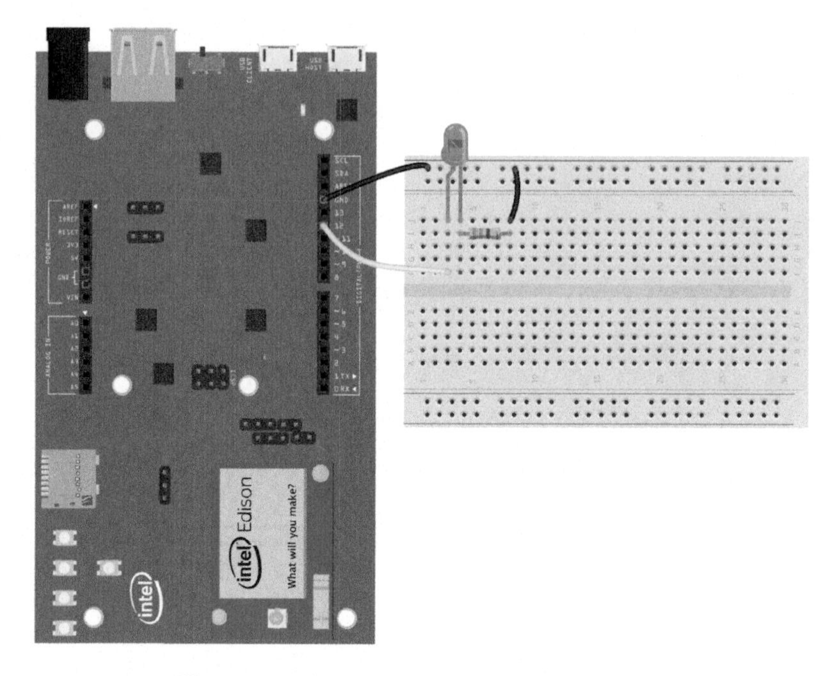

Figure 3-4. *The circuit layout for the blink example*[1]

If you wired the circuit exactly like Figure 3-4, then your LED is connected to pin 12. Modify your blink code to account for this change. Change:

```
int ledPin = 13;
```

to:

```
int ledPin = 12;
```

Upload it and you should see the LED on your breadboard blink (and the one wired to pin 13 stop blinking). You could actually use any of the digital pins (1-13) for this example, so long as the wired pin matches the Arduino software programming. The choice of pin 12 was completely arbitrary.

1 *All circuit layouts and circuit diagrams in this book were created using the amazing Fritzing software (http://fritzing.org/home/).*

In the previous examples, we used the onboard circuit connected to pin 13, and we didn't worry about adding a resistor or connecting the LED to ground. This is because all of this is already built in to the circuit on the Edison board.

Physically, what's happening? When you use `digitalWrite` to set pin 12 to `HIGH`, you're setting the source to five volts. This voltage is higher than the ground, which causes current to flow through the LED and the LED to emit light. When you set `digitalWrite` to `LOW`, you're setting the source to zero volts. This is equal to the voltage at ground, so the current stops flowing and the LED stops emitting light.

The resistor is added to the circuit to intentionally increase the load. In general, increasing the load in a circuit decreases the current flowing through it. In this case, we want a lower current to keep from damaging our LED by having it glow too hot or too bright.

It's worth taking a closer look to see what, mathematically, is happening here. There is a governing principle for voltage known as Ohm's law, which states that the voltage drop (V) across a conductor is proportional to the current (I) by the resistance (R), or $V = IR$. Therefore, if you were to wire a circuit as shown in Figure 3-5, the only way to get from the high-voltage source to the low-voltage sink would be to drop the total voltage across the resistor (assuming the wires have 0 resistance). If the voltage supply were equal to 5 V, and this single resistor had a value of 1 kohm, then the current would be $I = V/R = 5V / 1{,}000\ ohms = 5\ mA$ flowing through the circuit.

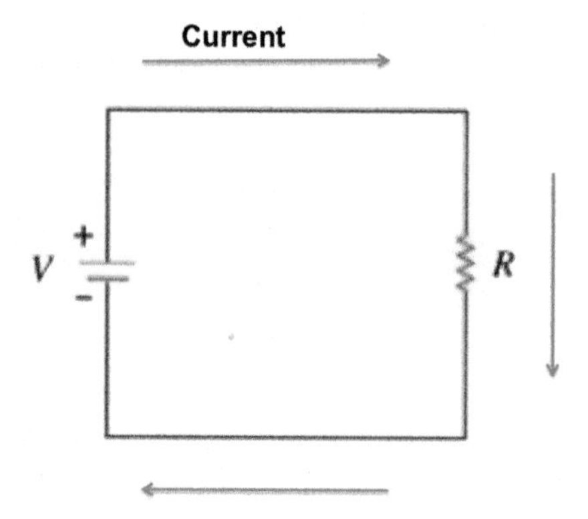

Figure 3-5. *Circuit diagram of a simple circuit with only a voltage source and a single resistor*

However, your LED circuit is a bit different. There's an additional diode component, as shown in Figure 3-6. Diodes are not like resistors. Resistors reduce voltage according to Ohm's law, whereas diodes always drop a fixed voltage inherent to their semiconductor properties. Kirchhoff's second law (no more laws after this, I promise!) states that the total voltage drop around any closed loop is equal to the input voltage.[2]. For our LED circuit, this means that the voltage dropped across the diode plus the voltage dropped across the resistor must equal the source voltage of 5V.

2 I'm paraphrasing here and simplifying quite a bit. For a full description, see the Wikipedia page (*https://en.wikipedia.org/wiki/Kirchhoff%27s_circuit_laws*).

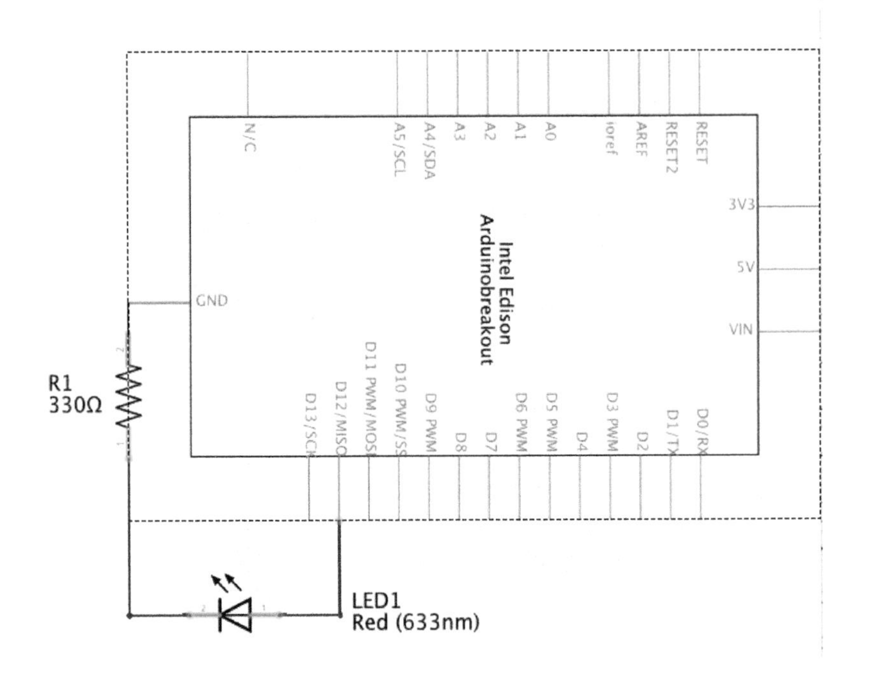

Figure 3-6. *Circuit diagram of a simple circuit with only a voltage source and a single resistor*

Why bring this up? Aside from just knowing how circuits work, this information helps you to calculate the proper resistor to use in your LED circuits. You'll have to check the ratings on your specific LEDs, but most want a current in the range of 15 mA and induce a voltage drop of approximately 2V. When pin 12 is set to `HIGH` we have an input voltage of 5V, which must be dropped across our LED and resistor:

```
Input Voltage = (LED Drop) + (Resistor Drop)
5 V = (2 V) + (0.015 A)*R
3 V = (0.015 A)*R
R = 200 ohms
```

Notice that it won't matter in which order you arrange your circuit. Whether you drop the voltage first across your resistor or first across your LED, the total voltage drop will remain the same.

In case you're wondering, there are also calculators online (*http://led.linear1.org/1led.wiz*) that can do this for you so long as you know your specs.

Digital Input: Adding a Button

Your LED circuit uses an LED as an output device, since the LED takes a signal from Edison and displays it to the world in the form of light. Conversely, an input device or sensor takes a signal from the outside world and transmits it into your Edison. One example of such a device is a push button. A push button is a digital device because it exists in one of two states: up or down. You'll now build on your blink circuit and sketch by adding a button that toggles the LED on and off instead of having it blink continuously.

To add the button interaction to your sketch, you'll need to declare two variables: `buttonPin`, the pin to which the button is connected, and `buttonState`, which monitors the status your button:

```
int buttonPin = 2; // The choice of pin 2 is arbitrary
int buttonState = 0;  // Initialize the button state
                      // to zero = button "up"
```

Next, in the `setup` function, initialize the button pin as an input:

```
pinMode(buttonPin, INPUT);
```

Finally, in the `loop`, write code to constantly monitor the state of your button:

```
buttonState = digitalRead(buttonPin);
```

Your entire sketch should look like this:

```
int ledPin = 12;  // LED pin
int delayTime = 1000;  // blink timing
int buttonPin = 2; // The choice of 8 is arbitrary
int buttonState = 0;  // Initialize the button state to zero =
button "up"

/* Function to blink the LED */
void blinkIt() {
  digitalWrite(ledPin, HIGH);
  delay(delayTime);
  digitalWrite(ledPin, LOW);
```

```
    delay(delayTime);
  }

  void setup() {
    pinMode(ledPin, OUTPUT);
    pinMode(buttonPin, INPUT);
  }

  void loop() {
    buttonState = digitalRead(buttonPin);
    blinkIt();
  }
```

Create the circuit shown in Figure 3-7 to match your new sketch.

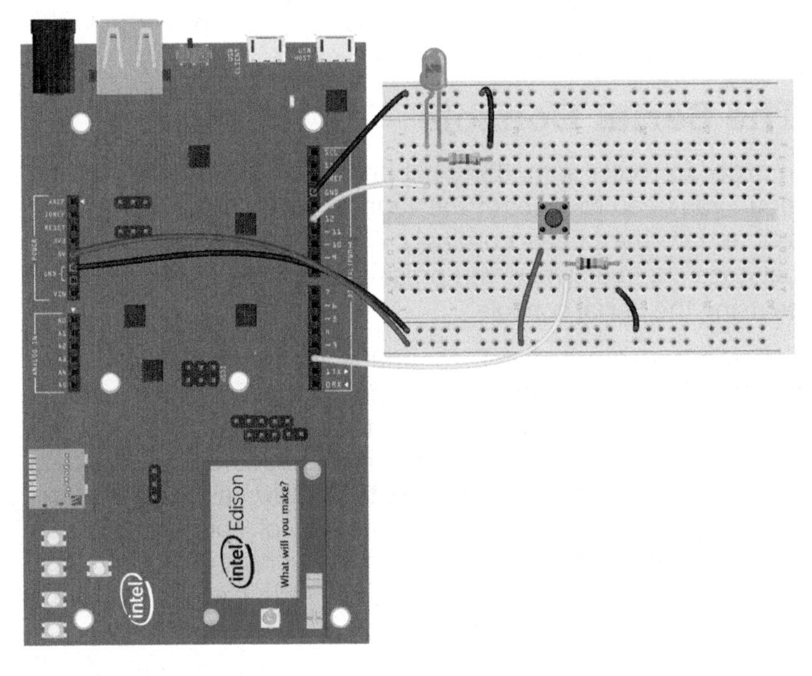

Figure 3-7. *Circuit layout for the button example*

Where to Begin

Some people find it easier to start by designing the circuit as hardware, either as a sketch or on the board, and then making the software match. Others find it easier to start by designing in the software and then wiring up the hardware accordingly. It's a matter of personal preference; there is no right or wrong way to do it.

Once you've built the circuit, upload your sketch to your Edison. If you press on the button, you might be sorely disappointed when nothing happens.

The Serial Console

In order to make something happen in response to the button input, you need to monitor the variable `buttonState` for changes and update the LED accordingly. First, let's check out how `buttonState` changes when you press the button. To do this, you'll be using the Serial console.

Serial is one protocol for communication between electronic devices. The two ends of the communication line can pass information back and forth, but first they have to agree on how quickly they pass the electronic signals that constitute the data. The rate at which they communicate is known as the *baud rate*. In fact, you've already used this method of communication in Chapter 1 and Chapter 2 when you used screen or Putty to connect to your Edison. In both cases, you specified a baud rate of 115200 when making the connection.

To enable the serial console in the Arduino IDE, add the following line to the setup function:

```
Serial.begin(9600);  // Set baud rate to 9600
```

Now add a line to the loop function that prints out the state of the button to the serial console. This line should come directly after you read the button state:

```
buttonState = digitalRead(buttonPin);
Serial.println(buttonState);  // print the button state to the
                              // console
```

This will print a line with the button state to the console every time we go through the loop function. Currently, this code is still blinking the LED. Each blink of the LED takes two seconds, which is too long to wait to see the button output. Delete the following line from the `loop` function:

```
blinkIt();
```

You can leave the actual blinkIt function in the code without causing any harm, but if you're a neat freak, feel free to delete that now as well.

The `loop` function will now run as fast as it possibly can, as there are no longer any delays programmed in. Edison will spit out data too quickly for us to process, so let's program in a small delay after the serial command:

```
delay(100);  // Small delay so that our eyes can catch up
             // to the console
```

Perfect! Upload your new code.

The Arduino IDE ships with a serial monitor that allows you to view the serial output from your Edison. In the main toolbar, go to Tools → Serial Monitor and it will pop up on the screen. You can also use the keyboard shortcuts CMD+Shift+M on a Mac or Ctrl+Shift+M on Windows or Linux to open the serial monitor.

Nonsensical Output

If ever your serial console is printing a string of seemingly random garbage characters, check the baud rate. When your computer and Edison are trying to communicate data at different rates, the messages get scrambled in translation. The rate you set using `Serial.begin()` must match the rate in the Serial Monitor drop-down window.

Since the button is currently in the "up" state, the serial monitor is printing a stream of zeros every 100 ms. When you press the button down, these zeros change to ones, because pin 2 is now receiving an input signal. The circuit diagram in Figure 3-8 sheds light on why this is so.

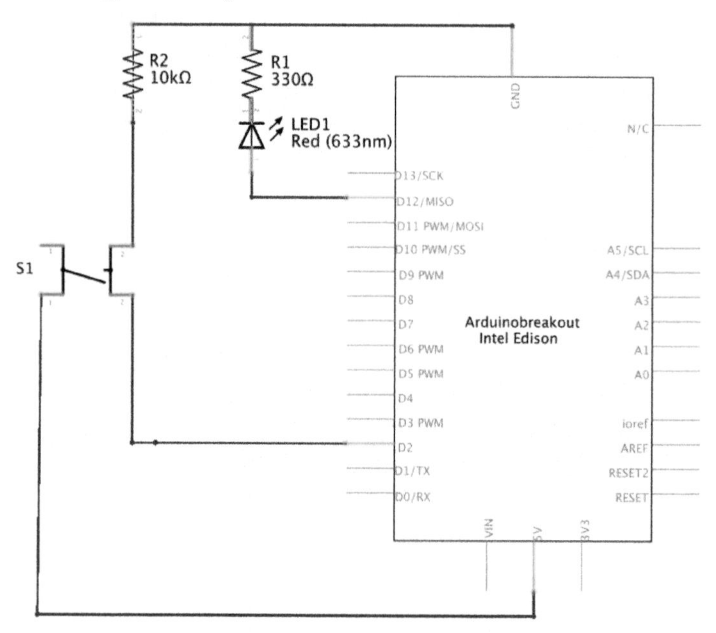

Figure 3-8. *Circuit diagram for the button example*

Before the button is pressed, the loop connecting 5V to digital pin 2 is left open or disconnected. At this time, pin 2 receives a zero voltage signal as an input. When you press the button, you're physically connecting the circuit loop, so digital pin 2 suddenly reads the 5V signal at this point in the loop. Letting the button go opens the circuit again and the reading goes back to zero.

The resistor running from the button to ground is known as a *pull-down resistor* because it is connected to ground. This resistor holds the signal reading at 0 V when nothing else is connected. More common are actually *pull-up resistors*, which hold the signal at high when nothing else is connected. For more information, check out the SparkFun tutorial (*https://learn.sparkfun.com/tutorials/pull-up-resistors*).

Unlike your LED circuit, the order of components in your button circuit actually matters. If you arrange your circuit so that the current passes through the resistor before it hits the button, then the resistor, as the only dissipative element in the ciruit, will have already dropped the voltage back down to 0 V when it's read by Edison. Try it. Whether you press the button or not, the serial console will always print 0s.

From this example, you can see why the serial console is a super useful debugging tool. If ever you build a circuit that isn't behaving the way you think it should, printing out the measured values along the path of the circuit can help you track down where the problem(s) resides.

Toggling the LED

Now let's change your code so that every time the button state goes from 0 to 1 (i.e., from pushing the button), your LED toggles on/off. You no longer need the `delayTime` variable, but you do need to track the state of the LED to toggle it. So, change the `delayTime` variable to the variable `ledState`, setting it initially to 0 since the LED will start off:

```
int ledState = 0;  // blink timing changed to ledState
```

You should only toggle the LED if the value of `buttonState` was 0 before performing `digitalRead` and 1 after. Every time through the loop, we need to check if this is true. Fortunately, programming languages come with a built-in function for just this purpose, if. *If* introduces a block of code that is only run if the condition it's monitoring is true. The following is an example:

```
if (ledPin == 13) {
        Serial.println("LED pin is 13!");
}
```

If you had declared pin 13 as your `ledPin`, the line "LED pin is 13!" would print to the serial console. If you had declared it as any other pin, everything inside the curly brackets would simply be skipped. If you want a certain code to execute if the condition is not met, you can use the *else* statement:

```
if (ledPin == 13) {
        Serial.println("LED pin is 13!");
}
```

```
else {
        Serial.println("LED is not on pin 13.");
}
```

In the preceeding example, if you declared the `ledPin` as any pin other than 13, the line, "LED is not on pin 13." would print to the serial console.

For your specific circuit and purpose, you want to make sure two different conditions are met. You can check for multiple conditions by linking them with `&&` symbols. As the very first thing in the `loop` function, add the following:

```
if ((buttonState == 0) && (digitalRead(buttonPin) == 1)) {
    ledState = 1-ledState;    // switch the ledState
    digitalWrite(ledPin, ledState); //  Write it to the led
}
```

Since it's the first command in the loop, you haven't updated the `buttonState` variable yet, so it still reflects the value from the last time through the `loop` function. If this value is 0 and `digitalRead` returns a 1, then you just pushed the button. The following command then toggles the value of `ledState` by subtracting it from 1:

```
ledState = 1-ledState;    // switch the ledState
```

If it was 0, it becomes 1. If it was 1, then it becomes 0. Since you just switched the value, you can write this new value to the `ledPin` using `digitalWrite`.

Upload your sketch and try it. The LED should switch on/off every time you push the button. For reference, your entire sketch should look something like this:

```
int ledPin = 12;  // LED pin
int ledState = 0;  // blink timing changed to ledState
int buttonPin = 2; // The choice of 8 is arbitrary
int buttonState = 0;  // Initialize the button state to 0.
Zero is button "up"

void setup() {
  pinMode(ledPin, OUTPUT);
  pinMode(buttonPin, INPUT);
  Serial.begin(9600);
}
```

```
void loop() {
  if ((buttonState == 0) && (digitalRead(buttonPin) == 1)) {
    ledState = 1-ledState;
    digitalWrite(ledPin, ledState);
  }
  buttonState = digitalRead(buttonPin);
  Serial.println(buttonState);  // print the button state to
                                // the console
  delay(100);
}
```

You might notice that the `delay(100)` is not really necessary anymore, since you're not really looking at the serial console for debugging. If you remove this `delay(100)`, rerun the example, and play with it for a while, you'll notice that the circuit is slightly less accurate at picking up your button presses. What's happening?

It turns out that microcontrollers read signals much more quickly than a person is able to give a mechanical input. In the instant you press the button, the signal fluctuates, or bounces, back and forth fairly rapidly before settling into the high or low state. As opposed to adding a 100 ms delay (which is more time than it takes for the signal to settle), a common fix is to add what is known as *debouncing*, which can be incorporated on either the hardware or software side. A good lesson and some code on software debouncing can be found on the Arduino site (*http://playground.arduino.cc/Learning/SoftwareDebounce*).

Analog Output

The analog output on Edison differs from the digital output in that it can use a range of voltages between 0 and 5V in 255 different increments. Edison does not set these intermediate voltages directly; it uses a technique called *pulse-width modulation* to achieve an average voltage signal equal to the desired value. The output signal very quickly oscillates back and forth between 0 and 5V, as shown in Figure 3-9. Pulse-width modulation varies the ratio of the time spent in the high and low states. The signal oscillates quickly enough that connected devices experience the average voltage instead of the rapidly oscillating high or low.

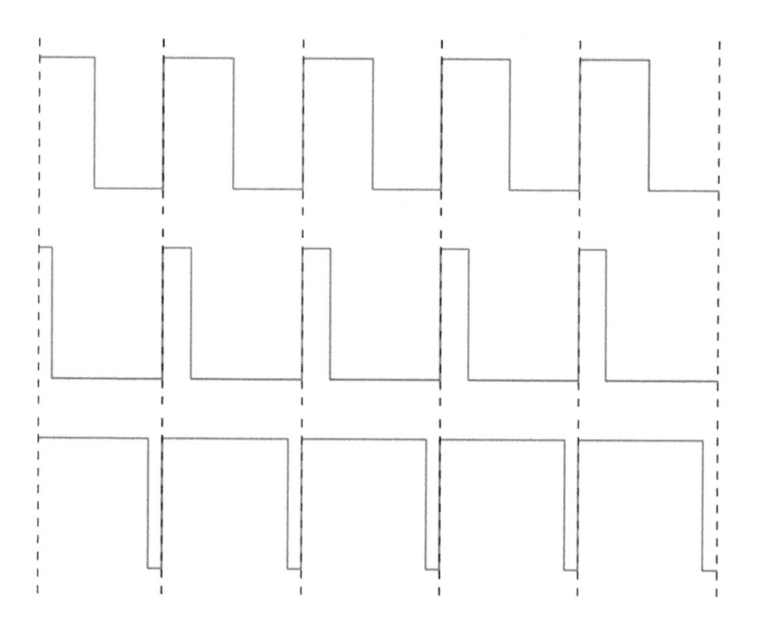

Figure 3-9. *An example of pulse-width modulation giving average voltage signals equal to 50, 20, and 90 percent of the maximum value*

The function for specifying the analog output voltage is, unsurprisingly, `analogWrite`. Specifying a value of 255 to `analogWrite` keeps the pulses at 5V 100% of the time, yielding a 5V signal. Specifying a value of 127 keeps the pulses at 5V for 50% of the time, yielding a 2.5V signal. The percentage of time that the output signal spends in its high-voltage state is known as the *duty cycle*.

The PWM-enabled pins on your Edison are the digital pins denoted with a ~ next to the pin number. Because of hardware limitations, only four of these pins can be enabled at a time. By default, the active PWM pins are 3, 5, 6, and 9. For this exercise, move your LED circuit from the blink example over to pin 9 by moving the yellow wire in Figure 3-4 3 pins down from pin 12. You're going to make this LED glow brighter as the result of increasing the analog voltage over time.

First, declare the variables you'll need before the `setup` function: the LED pin number, the brightness, the `delayTime`, and the step size:

```
int ledPinAnalog = 9; // LED pin, selected arbitrarily
int delayTime = 100;  // how quickly brightness
                      // will increase/decrease
int brightness = 0;   // Initial brightness of 0, scale 0-255
int stepSize = 10;    // Amount to increase the brightness
                      // each time
```

For this example, you'll leave the `setup` function blank; analog output does not require you to initialize the pins as digital I/O did. Don't delete the function, however, or your code will throw an error when compiling.

Within the `loop` function, you'll first increase the brightness by the step size and then write that increased value to pin 9 using `AnalogWrite`. This will increase the voltage on pin 9 each time through the loop, making the LED glow brighter over time. In order to observe these changes with the naked eye, you'll also have to program in a small delay. All in all, your loop function should look something like this:

```
void loop() {
  brightness += stepSize;
  analogWrite(ledPinAnalog, brightness);
  delay(delayTime);
}
```

The line:

```
brightness += stepSize;
```

is short for:

```
brightness = brightness + stepSize;
```

And is a pretty handy coding trick. You can use this trick with any of the arithmetic operations (*, /, +, and -) and a range of other operators others as well.

Upload this sketch. You'll notice that the LED gets brighter at first but then stops increasing at some maximum brightness value. This is because `AnalogWrite` expects a value between 0 and 255, and the preceeding code continues to increase the brightness variable indefinitely. Once the brightness exceeds 255, Edison treats it as a 255, and the LED simply glows at max brightness.

A simple fix is to program in a check on the **brightness** variable. If the value exceeds 255, set the **brightness** back to zero or subtract 255 from the value, forcing it back into the acceptable range. Applying this fix, your **loop** function should look like this:

```
void loop() {
  brightness += stepSize;
  if (brightness > 255) {
    brightness -= 255;
  }
  analogWrite(ledPinAnalog, brightness);
  delay(delayTime);
}
```

Upload this sketch, and you'll notice that the LED now resets after it becomes maximally bright. Play with the **delayTime** and **stepSize** to see what looks the best. See if you can program your LED to dim instead of brighten, or brighten and then dim back down.

Analog Input

Intel Edison has six pins dedicated to reading analog input signals, labeled A0 to A6. These pins read an analog, or continuous, signal from a sensor or other input device. The infinite number of values in a continuous signal are then discretized in the range of 0 (no voltage) to 1,023 (5V) by an onboard circuit known as an *analog-to-digital converter* (ADC).

Breakout Boards

Up until this point, all the exercise that you've done will work equally well on the mini breakout, assuming you've got your pins mapped correctly. This is, unfortunately, where that stops. The Arduino breakout board has an onboard ADC, but the mini breakout does not. If you want to read analog signals with the mini breakout, you'll have to interface your own ADC. If you're working with the SparkFun Base Block, then you can purchase (*https://www.sparkfun.com/products/13046*) the ADC block to read analog signals.

The analog device you'll be using for this example is a *potentiometer*. Unlike the push button that only returned one of two states, the potentiometer is a knob can be turned continuously to provide a variable resistance. As you turn the knob, you change the resistance on either side of the center pin, making the signal at that point "closer" to either the 5V source or ground.

You can think of a potentiometer like two individual resistors with one fixed total resistance. This is shown in Figure 3-10. If you crank the potentiometer all the way in one direction, the R1 resistor will contain 100% of the total resistance. All voltage will drop across this resistor, and the Vout signal will read 0 volts. Reading the Vout signal using the `analogRead` function will return 0. When cranked all the way in the other direction, the R2 resistor will contain 100% of the total resistance and the voltage will not drop at all before reaching Vout. In this case, Vout will be 5 volts, and `analogRead` will return a maximum value of 1,023. By turning the potentiometer to the intermediary values of R1 and R2, any voltage in the range 0-5V can be obtained.

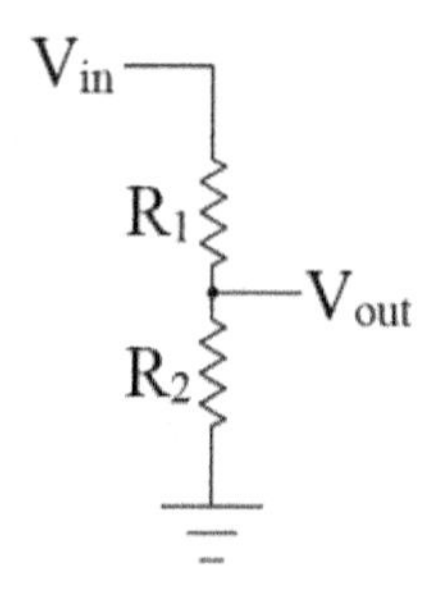

Figure 3-10. *A representation of a potentiometer as two separate resistors*

A schematic of the circuit is shown in Figure 3-11. You'll be using the analog signal from the potentiometer on pin A5 to control the brightness of the LED on pin 9. In this circuit, both pin A5 and pin 9 have been chosen arbitrarily; any combination of analog input pin and PWM output pin will work.

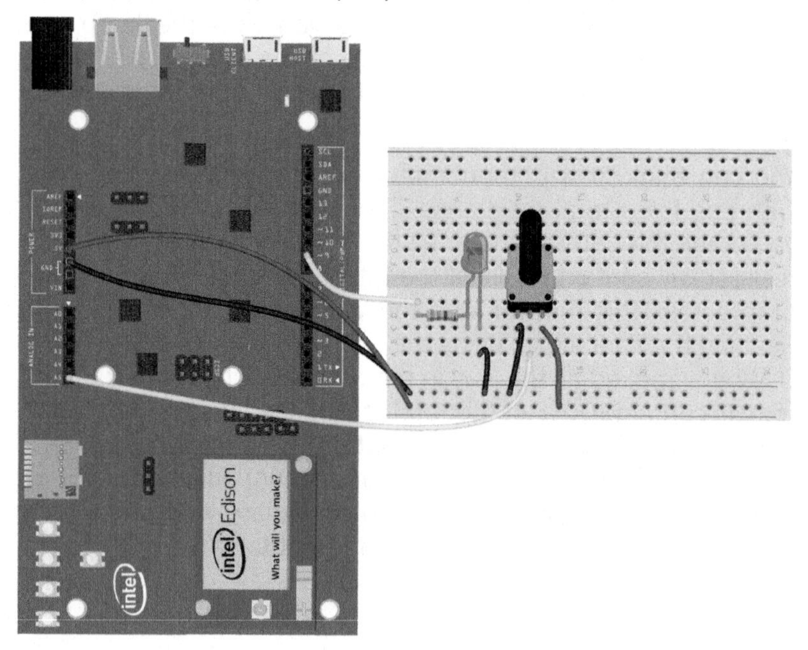

Figure 3-11. *Circuit layout for the potentiometer example*

You'll now pair the circuit with the Arduino code. First, declare the variables that you need: the pin numbers, the analog value read from the potentiometer, and the output brightness value for the LED:

```
int ledPinAnalog = 9;  // LED pin, selected arbitrarily
int potPinAnalog = A5;  // Analog in potentiometer pin,
                        // selected arbitrarily
int brightness = 0; // Initial brightness of 0, scale 0-255
int potValue = 0;  // Amount to increase the brightness each
time
```

As in the last example, because you're working with only analog pins, you can leave the **setup** function blank.

In the **loop** function, you'll need to do two things: read the analog value from pin A5 using **analogRead** and then write the analog value out to pin 9 using **analogWrite**:

```
void loop() {
  potValue = analogRead(potPinAnalog);
  analogWrite(ledPinAnalog, potValue);
}
```

Serial Output

You can also use the serial console to view the values read from the potentiometer or any other analog sensor. This is a great way to get a feel for the range of values that a sensor will return.

One small problem still remains. The **analogRead** function returns a value from 0 to 1,023, but the **analogWrite** function is capped at 255. In order to make the full range of light correspond to the full turning range of the potentiometer, you need to scale the potentiometer values to the 0 to 255 range. The Arduino IDE comes with a function for just this purpose: **map**. The **map** function takes five inputs—the unscaled value, the low and high of the original range, and the low and high of the desired range—and returns the scaled value as an output. Insert **map** into the **loop** function as follows:

```
void loop() {
  potValue = analogRead(potPinAnalog);
  brightness = map(potValue, 0, 1023, 0, 255);
  analogWrite(ledPinAnalog, brightness);
}
```

Upload your code, and you'll see that turning the potentiometer will adjust the LED brightness. Reading and writing analog signals opens up a wide range of devices to circuit programming. Outputs can be as varied as lights, and motors and inputs span all kinds of sensors: light, temperature, pressure, and even sound.

I2C Accelerometer

I2C (pronounced "I-squared-C" or "I-two-C" or "I-I-C") is an acronym for *inter-integrated circuit*. It is a protocol developed by Philips Semiconductors in the late 1970s to simplify the lines that travel among various circuit components. From the previous examples, it's probably already apparent that complicated circuits will lead to a ton of wired connections and a rather large number of I/Os if only analog and digital signals are used. The I2C protocol reduces the number of communication wires to two by allowing multiple peripherals to communicate on the same lines. In fact, you can wire up as many as 112 devices on just these two I2C lines.

Today, I2C has become so prevalent that there are literally hundreds of off-the-shelf components that use I2C to communicate with your Edison. These range from memory chips to LCD controllers and even amplifiers. For this example, we'll be using an accelerometer.

The two lines that I2C requires are a data line (SDA) and a clock line (SCL). You can connect these to either the pins labeled SDA and SCL on the digital side of the board or pins A4 (SDA) and A5 (SCL) on the analog side. There are only two other lines that need to be connected: Vin on the accelerometer to 5V on the Edison board and GND on the accelerometer to any ground pin on the Edison board. Your wiring should look like Figure 3-12.

Figure 3-12. *Wiring image for the I2C accelerometer example*

In order to communicate with our I2C accelerometer, you need to load an Arduino library with a specific set of communication functions. This library is called `Wire` and can be loaded by adding the following to the very top of your Arduino code:

```
#include <Wire.h>
```

In this example, all you do is confirm that the accelerometer is actually the device it's supposed to be. To do so, you'll have to transmit information to the I2C device, read back from it, and confirm that it gives the response you expect. To transmit information to the device, you'll have to send at least two bytes: the unique address of the accelerometer, in this case 0x1D, and the *register* of the information you're reading or writing. A register is the unique identifier of information on the device itself. Declare these two variables before the `setup` function. They are declared as constants because they will not change anywhere in the code:

```
const unsigned char _i2caddr = 0x1D;
const unsigned char MMA8451_REG_WHOAMI = 0x0D;
```

Also declare a one-byte variable for the ID of the device that we'll read back:

```
unsigned char deviceid;
```

Hex Format

0x1D and 0x0D are numbers declared in hexadecimal, or base 16, format. A hexadecimal number is just a single byte integer in the range 0 to 255 written in a different format. It is the most common format in which to see device registers, addresses, and commands written.

Within the `setup` function, initialize serial output and then initialize communication with the device by calling `begin`:

```
Serial.begin(9600);
Wire.begin();
```

Next, use the wire library to send the device address over the data line. This notifies the accelerometer that information is coming:

```
Wire.beginTransmission(_i2caddr);
```

Next, write the `WHOAMI` register to the device and request 1 byte back as a response:

```
Wire.write(MMA8451_REG_WHOAMI);
Wire.endTransmission(false); // don't end transmission
                             // on MMA8451
Wire.requestFrom(_i2caddr, 1);
```

Finally, perform a series of checks using `if` statements:

```
if (Wire.available())   ❶
  {
    deviceid = Wire.read();   ❸
    /* Check connection */
    if (deviceid != 0x1A)  ❹
    {
      /* No MMA8451 detected ... return false */
      Serial.println(deviceid, HEX);
    }
    else  ❺
```

```
    {
      Serial.println("MMA8541 Detected!");
    }
  }
  else {
    Serial.println("Device not connected.");  ❷
  }
```

❶ Check if the device is available at all.

❷ If the device is not available, print an error message, "Device not connected," to the serial console.

❸ If the device is connected, read the byte you requested.

❹ If it does not return 0x1A, print the returned byte to the serial console.

❺ If it does return as 0x1A as expected, then print the message, "MMA8541 was detected."

The entire sketch is shown here:

```
const unsigned char _i2caddr = 0x1D;
const unsigned char MMA8451_REG_WHOAMI = 0x0D;
unsigned char deviceid;

void setup() {

  Serial.begin(9600);

  // put your setup code here, to run once:
  Wire.begin();

  Wire.beginTransmission(_i2caddr);
  Wire.write(MMA8451_REG_WHOAMI);
  Wire.endTransmission(false); // MMA8451 + friends use
                               // repeated start!!
  Wire.requestFrom(_i2caddr, 1);
  if (Wire.available())
  {
    deviceid = Wire.read();
    /* Check connection */
    if (deviceid != 0x1A)
    {
      /* No MMA8451 detected ... return false */
      Serial.println(deviceid, HEX);
    }
```

```
      else
      {
        Serial.println("MMA8541 Detected!");
      }
    }
    else {
      Serial.println("Device not connected.");
    }

}

void loop() {
  // NOTHING!
}
```

When you run this sketch and open the serial console, you should see "MMA8541 Detected!" If not, check your wiring and make sure everything is connected properly.

You might be wondering if you'll have to write all the programming for the MMA8541 this way. The answer is no. Arduino has a very active user community that almost always writes libraries to interface with the Arduino-compatible hardware. It is extremely rare that you'll need to write any I2C calls from scratch as you just did. Specifically, the MMA8451 has an associated library that you can download from GitHub (*https://github.com/adafruit/Adafruit_MMA8451_Library*) as a ZIP file. Unpack the file and remove the -master from the folder name (Arduino won't read the -). Close the Arduino IDE and move this folder into the libraries directory within your main Arduino folder. Repeat the same process with the Adafruit sensor library (*https://github.com/adafruit/Adafruit_Sensor*). The MMA8451 code depends on this library and won't run without it.

Now, restart the Arduino IDE, which will automatically source the new library files. Note that if you add new libraries while the Arduino IDE is still open, they will *not* be sourced. You have to close and reopen the IDE for the software to recognize them.

Go to File → Examples → Adafruit_MMA8451_Library and select the *MMA8451demo* file. Upload this sketch to your Edison and open the serial console when the upload finishes. The output should look like Figure 3-13.

```
Portrait Up Front

X:      -140    Y:      724     Z:      3878
X:      -0.03   Y:      0.19    Z:      0.99    m/s^2
Portrait Up Front

X:      -144    Y:      834     Z:      4000
X:      -0.04   Y:      0.20    Z:      0.98    m/s^2
Portrait Up Front

X:      -136    Y:      786     Z:      4000
X:      -0.03   Y:      0.21    Z:      0.99    m/s^2
Portrait Up Front
```

Figure 3-13. *Serial output of the MMA8451 accelerometer code*

Despite saying m/s^2, the units of the second line are actually in units of gravity (G), where 1 G is equal to 9.8 m/s^2. If your accelerometer is sitting flat on a surface, then gravity in the z-direction should be close to 1, since it feels the earth's gravity down, and the other two axes should be close to 0, since they experience very little gravity in those transverse directions. Using all three axes values, the accelerometer can calculate what orientation it's in. This is listed below the values (here as "Portrait Up Front"). Move your accelerometer around and see how these values and orientations change. Tap or lightly drop your accelerometer and see what sort of readings you get. Accelerometers are incredibly useful devices; they're used for everything from measuring car crash impact to orienting your smartphone display. Now you have your own hardware and tools for measuring acceleration and orientations.

SPI Screen

SPI, short for *serial peripheral interface*, is another standard protocol that is used to connect multiple peripherals to the same data lines. SPI operates using a slave-master architecture, meaning one device has full control over all peripherals. In the vast majority of cases, Edison will be your master device, controlling the slave devices that are attached to the SPI lines.

Unlike I2C, each SPI device requires four connections, one of which is not shared (thus each peripheral requires an additional pin):

Master Out Slave In (MOSI, DOUT)
: A shared line on which data is sent from the master to the slave. On the Arduino breakout, this is pin 11.

Master In Slave Out (MISO, DIN)
: A shared line on which data is sent from the slave device to the master. On the Arduino breakout, this is pin 12.

Clock (CLK, SCLK, or SCK)
: A shared line that provides the serial clock time from the master. On the Arduino breakout, this is pin 13.

Slave Select (SS, CS, nSS, or nCS)
: A digital output from the master stating whether this device is in use. Any digital pin can be used, and the default standard is that the device is active when this signal is low.

To transmit information over SPI, the SS pin is first set to low. Since each peripheral has its own SS pin, this selects the peripheral attached to that specific line for communication. Next, during each clock cycle, a bit is transmitted from master to slave over MOSI and then from slave to master over MISO. After the signal has been fully transmitted, the SS pin is set back to high, deactivating the communication with that particular device.

The device you'll be using for SPI communication is the Adafruit 2.8" TFT Touch Screen Display. If you purchased the shield version, plug it in by recessing it into the Arduino breakout in its natural orientation (the direction in which the ICSP pins slide into place). If you're instead using the breakout, wire it as shown on the second or third image on this page (*https://www.adafruit.com/products/1770*). The image shows an Arduino, but the pins are the same on the Edison Arduino breakout. Note that the first image when loading the page has a great many more connected wires. This is to utilize the full functionality—screen, resistive touch, and SD card—of the breakout board. For this example, we'll just be using the screen, so not all of the wires are necessary.

For this example, you'll be writing SPI instructions to the screen to adjust the display, which is driven by the ILI9341 controller. The ILI9341 controller is fairly popular as far as screens go, so you'll be able to control the screen of any device that utilizes it if you're interested in using something else.

Integrated Electronics

A lot of screens on Adafruit, Sparkfun, and other maker sights come with many integrated components: capacitive touch, resistive touch, SD card slots, etc. Before purchasing a screen, make sure to research these other components to ensure that libraries for Intel Edison are readily available for them. For example, Edison libraries exist for the STMPE610 resistive-touch element on the board we recommend for this chapter but are not as readily available for the FT2606 board on the capacitive touch model.

You'll need two libraries for this example: a custom library (*http://bit.ly/ILI9341*) for the ILI9341 for controlling the screen and the standard Adafruit GFX library (*https://github.com/adafruit/Adafruit-GFX-Library/archive/master.zip*) for drawing certain preprogrammed shapes, lines, and text. Download both of these, unzip them, and move them into your Arduino libraries folder as in the last example. Make sure to completely quit out of the Arduino IDE and restart it so it can properly source these files.

First, you'll do a simple example reading information from the device registers and marching "Hello World!" across the display screen. At the very top of a blank sketch, add the following:

```
#include "SPI.h"
#include "Adafruit_GFX.h"      ❶
#include "Adafruit_ILI9341.h"

// Default SPI pins    ❷
#define TFT_DC 9
#define TFT_CS 10
```

```
#define TFT_MOSI 11
#define TFT_MISO 12
#define TFT_CLK 13

int row = 0;    ❸
int col = 0;

Adafruit_ILI9341 tft = Adafruit_ILI9341(TFT_CS, TFT_DC);    ❹
```

❶ Import the libraries that you'll need for communication.

❷ Define the SPI communication pins as constants (as opposed to the variables that you've been using for other sketches).

❸ Declare integers for the row and column positions.

❹ Initialize the touchscreen object using the Adafruit_ILI9341 library functions.

#define Versus Variable Declaration

Using *#define* assigns a name to a constant value before the program is compiled, as opposed to declaring variables, which happens at compile time. Most microcontroller boards have low on-chip memory, and defining constants don't occupy their precious program memory space on Arduino. Edison, however, has heaps of memory, so either defining or declaring should be fine in pretty much all instances.

Next, add the following as your setup function:

```
void setup() {
  Serial.begin(9600);      ❶
  Serial.println("ILI9341 Test!");

  // Begin tft object, init'ing SPI communiction
  tft.begin();    ❷

  // Read some diagnostic registers from the device
  uint8_t x = tft.readcommand8(ILI9341_RDMODE);    ❸
  Serial.print("Display Power Mode: 0x");
  Serial.println(x, HEX);
```

```
    x = tft.readcommand8(ILI9341_RDPIXFMT);
    Serial.print("Pixel Format: 0x"); Serial.println(x, HEX);

    x = tft.readcommand8(ILI9341_RDSELFDIAG);
    Serial.print("Self Diagnostic: 0x"); Serial.println(x, HEX);

  }
```

❶ Initialize serial communication and print a debug message to the screen.

❷ Begin the information exchange with the touch screen.

❸ Read some diagnostic information from the registers on the device, such as power mode, pixel format, and diagnostic.

Finally, you'll program your actual drawing in the `loop` function:

```
void loop() {
  // put your main code here, to run repeatedly:
  tft.fillScreen(ILI9341_BLACK);   ❶
  tft.setTextColor(ILI9341_WHITE);    ❷
  tft.setTextSize(1);
  tft.setCursor(row, col);   ❸
  tft.println("Hello World!");    ❹
  row += 10;    ❺
  col += 10;
  if (row > 240) { row -= 240; }    ❻
  if (col > 320) { col -= 320; }
}
```

❶ Fill the screen with black.

❷ Set the text properties, such as color and size.

❸ Set the cursor to the row and column location.

❹ Print the text, "Hello World!" to the screen.

❺ Add 10 to each of the row and column variables. By incrementing these variables each time through the loop, you make it appear as though the text is marching across and down the screen.

❻ To ensure that the text doesn't run off the bounds of the screen, use an `if` statement to check that the row is less than the screen width, 240 pixels, and the column is less than screen height, 320 pixels.

Upload the sketch to your Intel Edison. "Hello World!" will march across the screen and if you open the Serial console, you'll see the diagnostic information you requested in the `setup` function.

Finally, you can find several different example sketches in the *examples* folder of the *ILI_9341* directory. Open the *graphicstest.ino* file within the *graphicstest* directory. Upload this sketch to your Edison, and you'll see that it draws a wide array of shapes, forms, colors, and texts on the screen. You can use this file as a general template for drawing pretty much any image you can imagine.

Linux, C++, and the Arduino IDE

As mentioned in the Preface, programming Intel Edison using the Arduino IDE exposes all the libraries in the standard C++ distribution. One example that demonstrates this is the `system` function, which takes a single command-line command as an argument. This powerful function allows you to control the Linux command line from within the Arduino programming IDE. We can then redirect the command output to a file as you saw in Chapter 2, or you can pipe it to the Arduino serial console, which is simply the Linux path */dev/ttyGS0/*. Using the latter technique allows you to see if you're achieving the proper response from your system calls. Load the following code into a blank Arduino sketch:

```
void setup() {
  Serial.begin(9600);  ❶
  Serial.println("ifconfig test!");
}

void loop() {
  system("ping -c 2 google.com &> /dev/ttyGS0");    ❷
  delay(5000);    ❸
}
```

This code performs the following actions:

❶ Initializes the serial console.

❷ Uses the `system` call to ping Google. This call pings Google twice each time through the `loop` function and outputs the response to the serial console.

❸ Sleeps for five seconds.

Upload it to your Intel Edison, and open the serial console to see the output. If you copied correctly and your Edison is connected to the Internet, you should see a similar response in your serial console every five seconds:

```
PING google.com (173.194.46.100): 56 data bytes
64 bytes from 173.194.46.100: icmp_seq=0 ttl=54 time=29.389 ms
64 bytes from 173.194.46.100: icmp_seq=1 ttl=54 time=24.394 ms

--- google.com ping statistics ---
2 packets transmitted, 2 packets received, 0.0% packet loss
round-trip min/avg/max/stddev = 24.394/26.892/29.389/2.497 ms
```

Because Edison is available via WiFi, you can also use it to serve web content through the Arduino IDE. As you saw in Chapter 2, Edison automatically begins running a web server at *device name.local* or your Edison's IP address upon connection to the Internet. On the Edison filesystem, this web server is located in the */usr/lib/edison_config_tools/public/* directory. Any files placed in this directory can be accessed at *devicename.local/filename*.

Edison and the Internet

In Chapter 2 I mentioned that you can access Edison's web server so long as your host computer is connected to the same network. If you'd like to make your Edison accessible to the entire Internet, you'll need to assign it a reserved IP address and enable port forwarding on your router. This will allow Edison to connect using the same IP address each time it gets online and send/receive information through your router's firewall. Alternatively, a slightly more beginner-friendly and secure option is Weaved (*http://www.weaved.com*); an installer for Edison is available.

Wire up your Intel Edison as shown in Figure 3-7 and load or copy the following sketch into the Arduino IDE. This is mostly identical to the our original button sketch, except now each time you toggle the LED, you also set the text on your web page by echoing it to a file in the web-server directory:

```
int ledPin = 12;  // LED pin
int ledState = 0;  // blink timing changed to ledState
int buttonPin = 2; // The choice of 8 is arbitrary
int buttonState = 0;  // Initialize the button state to zero.
Zero is button “up”

void setup() {
  pinMode(ledPin, OUTPUT);
  pinMode(buttonPin, INPUT);
  Serial.begin(9600);
}

void loop() {
  if ((buttonState == 0) && (digitalRead(buttonPin) == 1)) {
    ledState = 1-ledState;
    digitalWrite(ledPin, ledState);
    system("echo '<html><head><title>LED State</title></head>'
> /usr/lib/edison_config_tools/public/LEDstatus.html");
    if (ledState == 0) {
        system("echo '<body><h2>LED is OFF!</h2></body></
html>' >> /usr/lib/edison_config_tools/public/LEDsta
tus.html");  }
    else {
        system("echo '<body><h2>LED is ON!</h2></body></html>'
>> /usr/lib/edison_config_tools/public/LEDstatus.html");  }
  }
  buttonState = digitalRead(buttonPin);
  Serial.println(buttonState);  // print the button state
                                // to the console
  delay(100);
}
```

Upload this sketch to your Intel Edison and open the page *device name.local/LEDstatus.html* in a browser window. As you change the state of the LED by pressing the button, refresh your browser to see these changes reflected on your website. In this way, you can use Intel Edison and the Arduino IDE to serve information to the Web from sensors around your home.

Troubleshooting

There are definitely some subtleties in moving the pre-existing Arduino libraries to the x86 architecture that Intel Edison (and Intel Galileo) run on. You might find that you run into issues when trying to move Arduino sketches and examples over to your Edison. This is especially true when using hardware libraries that are outside the common core; for example, the LCD display libraries.

There are two sites I found that list Galileo-compatible shields:

- Cooking Hacks (*http://bit.ly/galileo-hacks*)
- Open Hacks (*http://bit.ly/shields-list*)

If you want to use a shield on this list, then you're probably in the clear with Intel Edison. Just because a shield is not on the list doesn't mean that it won't work, but it's worth checking around on Google and the Intel Forums before purchasing.

Fortunately, Edison has a wide and growing user community that is working hard to enable more off-the-shelf Arduino programming each and every day. If you encounter problems, the first places you should check are the Edison Forums (*https://communities.intel.com/community/tech/edison/content*) and the Edison Troubleshooting Guide (*http://bit.ly/edison-guide*).

Going Further

Now that you've mastered some basic examples and circuits, there's so much more to explore. The following resources provide a good starting point for going further:

Make: Electronics (http://www.makershed.com/products/make-electronics-book)
: Maker Media's introductory guide to electrical components and circuitry. A great overview of modern maker electronics.

Arduino Cookbook (http://www.makershed.com/products/arduino-cookbook-2nd-edition)
: A great guide, for beginners and advanced users, to programming with Arduino.

Learn C++ (http://www.learncpp.com/)

A very comprehensive series of tutorials in learning how to program C++. Learning C++ is useful not only for leveraging the full power of the programming language in the Arduino IDE, but also for writing standalone applications that can run on your Intel Edison. Slightly more to come on this in Chapter 7.

4/Programming in Python

The first programming language that I ever learned was Java. This was followed by C and then shortly after by Fortran. In college, a professor saw these on my resume and told me to just list Python. "It's so easy by comparison," he claimed, "that you can just learn it in a day if anyone asks." I think he might have been right. My college has since changed their introductory computer science course to Programming in Python.

Introduction

Python is the language of choice for many developers and even more so for many hackers and makers. It's widely regarded as a great first programming language for many reasons:

Interpreted language
: Python is an interpreted language, meaning you can execute scripts directly through an interpreter instead of first compiling them into machine code. This adds the benefit of not having to explicitly declare variable types as we did in the Arduino IDE. The interpreter figures out for you which variables are ints, floats, etc.

The interactive shell

The Python interpreter can execute full Python scripts directly or as individual commands at an interactive shell. The shell presents a powerful way to develop scripts line by line without the hassle of writing a whole program around it.

The user community

The developer community around Python is enormous and incredibly active. From microcontrollers to web development to big data, there are libraries and communities that exist to help you get started and move you along.

In the last chapter, we used the Arduino IDE to control our circuits. In this chapter, we'll use the Python programming language to perform those same exercises and even add a few more.

The only additional equipment that you'll need is a smartphone or tablet—either iOS or Android—that is BLE enabled. It's best if you can get your hands on an Android, as one of the exercises in this chapter ("Bluetooth-Controlled LED" on page 106) works for Android only.

"Hello, World" in Python

The best way to learn to program is to dive right in. Connect to the terminal on your Edison by using screen or Putty or ssh'ing in wirelessly ("Connecting" on page 16). Edison's Yocto Linux distribution comes prepackaged with Python, so open the interactive shell simply by typing `python` at the command prompt. As the shell opens, you'll be greeted with information about the Python version running on your Edison and a triple chevron (>>>), indicating that the interpreter is ready for your commands:

```
# python
Python 2.7.3 (default, Mar 31 2015, 14:39:49)
[GCC 4.8.2] on linux2
Type "help", "copyright", "credits" or "license" for more
information.
>>>
```

Python Distributions

Intel Edison comes preconfigured with Python 2.7.3. There are other, newer version of Python, Python 3.X, that are close to the same but differ from the 2.7.3 distribution in just a few ways. If ever you come across a Python tutorial online that isn't working for you, check the Python version number that it's written for. This is particularly applicable if you're doing anything that involves byte strings and/or byte arrays.

At the prompt, issue the following command and then press Enter:

```
>>> print "Hello, World!"
Hello, World!
```

The interpreter executes your command and spits back the response in the same window. "Hello, World," the quintessential first programming task, is so simple when compared to other languages, that it actually inspired an xkcd cartoon (*http://xkcd.com/353/*).

You can also use Python for simple and not-so-simple computations:

```
>>>  3+5
8
>>>  import math
>>>  math.sqrt(20374)
142.73752134600068
```

The `import` statement is like the `include` statement we used in the Arduino IDE. It tells the interpreter to include the code from certain libraries or modules, making them available for use. You can also declare variables and perform operations on them:

```
>>>  a = 2
>>>  a*math.atan(12)
2.9753101898129106
```

The interactive Python shell also supports tab completion on functions, libraries, and variables. If you type `math.` and then press the Tab key twice, you'll see all the functions available in the math module. Python also has a very useful built-in help statement:

```
>>> help("sum")
Help on built-in function sum in module __builtin__:

sum(...)
    sum(sequence[, start]) -> value

    Returns the sum of a sequence of numbers (NOT strings)
    plus the value of parameter 'start' (which defaults to
    0). When the sequence is empty, returns start.
```

If the help text is long, you can exit out of it by pressing the Enter key to scroll to the bottom or by pressing the q key.

Often you'll probably want to execute standalone Python scripts or programs, which can be done by creating a text file with the *.py* extension and then calling it from the command line. To do so, first exit out of the interactive shell by hitting Ctrl+D or using the command `exit()`.

Open a blank text file with the name *HelloWorld.py* using the text editor of your choice. Insert the following text, save the file, and close:

```
print "Hello, world!"
a = 10
a += 5
print a
```

Run the file and you'll see that it outputs the print statements to the shell:

```
# python HelloWorld.py
Hello, world!
15
```

For the remainder of this chapter, I'll be switching between the interactive shell and standalone scripting. You're also free to do the same, pasting shell scripts into text files and vice versa. If you'd like to run the standalone programs, simply save them with a *.py* extension and run as we did above.

Functions and Loops

As we saw in Chapter 3, functions are an incredibly useful coding tool. Functions in Python have a slightly different structure than in C++:

```
>>> def multiply(a,b):
...     return a*b
...
>>> a = multiply(10,6)
>>> a
60
```

Notice that you didn't have to declare types for inputs or returns, but you did have to indent the `return` statement within the `multiply` function. If you left this out, you'd receive an error message:

```
>>> def m(a,b):
... a*b
  File "<stdin>", line 2
    a*b
    ^
IndentationError: expected an indented block
```

This is one of the biggest hurdles for beginners in coding with Python. Its coding structure is determined almost entirely by colons and whitespace. Any line introducing a block of code, such as a function declaration or if statement, should be finished with a colon. Everything within that block of code, starting with the next line, should be indented. When the indentation stops, the interpreter knows that the code block is over. It's identical in purpose to surrounding our functions with the curly braces in the Arduino IDE.

Two constructs in Python will be important when you start translating your Arduino code into this new language: the *for* loop and the *while* loop. The for loop steps through its associated code block at each increment in the for statement;

```
for i in range(10):
    a = 2*i
    print a
```

The range function creates an array of 10 integers starting from 0:

```
>>> range(10)
[0, 1, 2, 3, 4, 5, 6, 7, 8, 9]
```

The code then steps through these values, multiplies each value by two, and then prints the result to the shell. You should see the even integers from 0 to 18 as the output of your `for` loop.

By contrast, the `while` loop executes its associated code so long as its condition is true. The following `while` loop performs the exact same action as the preceeding `for` loop:

```
i = 0
while (i < 10):
    a = 2*i
    print a
    i += 1
```

Changing this code to that in the following example will create a loop that runs indefinitely, since the condition will always be true. This is just like the `loop()` function in Arduino:

```
i = 0
while(True):
    a = 2*i
    print a
    i += 1
```

Use Ctrl+C to kill this code so that it doesn't run forever.

Installing Libraries

There are an incredible number of libraries that add functionality to the standard Python distribution. They encompass everything from machine learning to image and audio processing and even web streaming. In fact, the Instagram website was built with Python's Django library at its core.

If the standard Python distribution does not include a library you'd like to use, you can often install it using the package manager (*https://pypi.python.org/pypi/pip*) *pip*. To install pip, first use `wget` to grab the pip installation file *get-pip.py*:

```
# wget https://bootstrap.pypa.io/get-pip.py
```

Then, run the code you downloaded to install pip:

```
# python get-pip.py
```

A very useful library to have installed on Intel Edison is `pyserial` (*http://pyserial.sourceforge.net/*). Pyserial reads and writes serial data like the serial library we used in the Arduino IDE. You can install this library by calling pip at the command line:

```
# pip install pyserial
```

Similarly, should you wish to uninstall this package, just run the `uninstall` command:

```
# pip uninstall pyserial
```

Installing Python Packages

Although using pip and opkg to install Python packages are probably equally easy, it will often be much faster to use the opkg method. This is especially true of large libraries or modules. Packages for opkg have been prebuilt for Intel Edison, whereas pip needs to do the build at install time. It's best to use pip only in instances where the prebuilt packages do not exist in the opkg repository. Pyserial is one such package.

We'll explore many Python libraries later in the book; just know for now that if there's a specific functionality you're seeking, there's probably a Python library you can install that provides it. And in 99% of cases, installing it will be just as easy as it was for pyserial.

Blink in Python

Let's move our original Blink example into Python, blinking the onboard LED connected to pin 13. To make sure that your former Arduino sketches do not conflict with your new Python code, remove the stored Arduino sketches before beginning to program in Python:

```
# rm /sketch/*
```

Create the script `blink.py`, open it in a text editor, and add the following:

```
import mraa     ❶
import time

ledPin = 13          ❷
sleepTime = 0.2   # blink timing

x = mraa.Gpio(ledPin)
x.dir(mraa.DIR_OUT)

while True:        ❸
    x.write(1)
    time.sleep(sleepTime)
    x.write(0)
    time.sleep(sleepTime)
```

This code performs the following actions:

❶ The first two lines import the necessary libraries. The `mraa` library that we installed at the end of Chapter 2 gives Python the ability to interface with the Arduino board hardware. Time gives us access so the system clock.

❷ The next four lines declare variables for the LED pin and blink time and then initialize the GPIO pin as an output.

❸ The `while` loop writes the I/O pin, alternating to 1 (high) and 0 (low), sleeping for 0.2 seconds in between.

Save this program and run it, and you'll see that the onboard LED connected to pin 13 will blink until you kill the program with Ctrl+C.

Most code in this chapter makes pretty heavy use of Intel's mraa library (*https://github.com/intel-iot-devkit/mraa*). Libmraa is a library written in C/C++ with bindings to other languages such as JavaScript and Python. The goal is to allow easy interfaces with the I/Os on Galileo, Edison, and other platforms where port names and numberings automatically match the board that you are on.

MRAA Numbering

The mraa library has some really odd GPIO pin numbering schemes for the SparkFun Blocks and mini breakout board. If you're attempting to use one of these boards, the pin-mapping, along with the mraa documentation, can be found on GitHub (*http://bit.ly/mraa-github*).

Scripting at Bootup

One of the most common questions is how to make Python scripts start at boot-up as the Arduino sketches do. The easiest way is to create a script that calls your Python code on startup.

First, change into the */etc/init.d* directory. If this is your first time adding a start up script to Edison, then you'll have to create the directory first:

```
# mkdir /etc/init.d
# cd /etc/init.d
```

Using your text editor, create a shell script called `automatic.sh` and paste in the following contents:

```
#!/bin/sh
python /home/root/blink.py >> /dev/null 2>&1 &
```

The latter part of the command suppresses the output and runs the Python script in the background. Save the file and close it.

Change the permissions on the `automatic.sh` script to make it executable:

```
# chmod +x automatic.sh
```

Next, add the script as a system service with the default settings, meaning it will be executed last at bootup:

```
# update-rc.d automatic.sh defaults
```

If you now reboot the board, the script will run, and the blinking will automatically start. If you wish to remove this script from the startup process, change directories into */etc/init.d/* and issue the command:

```
# update-rc.d -f automatic.sh remove
```

Button-Controlled Blink

Digital inputs are handled by mraa as simply as digital outputs. To integrate a button into the blink circuit, recreate the circuit shown in Figure 3-7. The accompanying sketch that toggles the LED translates into Python as follows:

```
import mraa
import time

ledPin = 12         # LED pin number for mraa library
ledState = 0        # current state of led: off
buttonPin = 2       # button pin number for mraa library
buttonState = 0     # current state of button: not pressed

led = mraa.Gpio(ledPin)
led.dir(mraa.DIR_OUT)
button = mraa.Gpio(buttonPin)
button.dir(mraa.DIR_IN)

while True:
    if buttonState == 0 and button.read() == 1:
        ledState = 1 - button.read()
        led.write(ledState)
    buttonState = button.read()
    time.sleep(0.1)
```

This code should be fairly transparent based on the Arduino sketch you wrote before. In fact, most simple Arduino sketches will translate easily over to Python and should be extremely "readable."

Bluetooth-Controlled LED

A far more interesting exercise is to use a Bluetooth connection from your smartphone to toggle an LED. For this example, you'll use just the Arduino Breakout Board, toggling the onboard LED connected to pin 13. Unfortunately, exchanging serial data directly with Intel Edison will only work using an Android device, but the process of Bluetooth pairing will work with either iOS or Android. Whether you're on Android or iOS, pair your device now. Later, you'll do a BLE example that works with both systems, and you'll need to be paired.

Bluetooth Pairing

The first step is to enable Bluetooth on Edison with the `rfkill` command. You can check that Bluetooth is enabled by using the command `rfkill list`:

```
# rfkill unblock bluetooth
# rfkill list
0: phy0: wlan
        Soft blocked: no
        Hard blocked: no
1: brcmfmac-wifi: wlan
        Soft blocked: no
        Hard blocked: no
2: bcm43xx Bluetooth: bluetooth
        Soft blocked: no
        Hard blocked: no
3: hci0: bluetooth
        Soft blocked: no
        Hard blocked: no
```

Edison boots in a Bluetooth-disabled mode in order to save power. Every time you power cycle Edison, you'll need to issue the `rfkill` command to instantiate the Bluetooth again. If you're sure you want the Bluetooth automatically active with every boot, a good option is to add the command to a startup script.

Next, we have to pair our Bluetooth devices to each other. This step need only be done a single time for each device; the pairing will remain forever so long as you don't delete the paired Bluetooth devices from your phone or Intel Edison.

To pair, you'll use the command-line Bluetooth controller program, `bluetoothctl`. Open it by typing the name at the Edison command line:

```
# bluetoothctl
[NEW] Controller 98:4F:EE:02:F2:D2 edison [default]
[bluetooth]#
```

When bluetoothctl starts, the MAC address and name of your Edison's bluetooth will appear after the word "Controller." Issuing the command `help` will show you all the Bluetooth configuration options available in bluetoothctl:

```
[bluetooth]# help
Available commands:
  list                          List available controllers
  show [ctrl]                   Controller information
and more...
```

On your smartphone, enable Bluetooth discovery of your device by going to Settings → Bluetooth, confirming that Bluetooth is set to On, and setting the device to discoverable. Once the phone is discoverable, start Bluetooth scanning on your Intel Edison to find the phone's MAC address. This may take a few seconds of searching:

```
[bluetooth]# scan on
Discovery started
[CHG] Controller 98:4F:EE:02:F2:D2 Discovering: yes
[NEW] Device A8:66:7F:AB:AE:01 Stephanie Moyerman's Nexus 5
```

Once you have the MAC address of your phone, disable scanning and pair with your phone, making sure your phone is unlocked and active:

```
[bluetooth]# scan off
[CHG] Device A8:66:7F:AB:AE:01 RSSI is nil
[CHG] Controller 98:4F:EE:02:F2:D2 Discovering: no
Discovery stopped
[bluetooth]# pair A8:66:7F:AB:AE:01
Attempting to pair with A8:66:7F:AB:AE:01
```

After issuing the command, a pairing message will pop up on your phone. Select Pair so that the two devices can connect. Upon connection, the status will be indicated in bluetoothctl.

So that future connections can forgo this process, set your Intel Edison to discoverable and set the permissions to trust your smartphone indefinitiely:

```
[bluetooth]# discoverable on
[bluetooth]# trust A8:66:7F:AB:AE:01
```

Exit from bluetoothctl:

```
[bluetooth]# quit
[DEL] Controller 98:4F:EE:02:F2:D2 edison [default]
```

Exchanging Information

To handle the connection and information transfer between your phone and Edison, download the *SPP-loopback.py* made available from Intel's developers. SPP stands for *serial port profile*, which emulates a serial cable for wireless communication such as Bluetooth. We'll get to the loopback portion later:

```
# wget http://downloadmirror.intel.com/24909/eng/SPP-
loopback.py
```

Run this code in the background:

```
# python SPP-loopback.py &
```

This program runs indefinitely, performing a few different functions. First, it waits for a Bluetooth connection from the phone. When the connection happens, it opens a socket for data transmit and receive. The code then transmits any data received from the phone back again with the prefix `"looping back:"`. Finally, when the phone disconnects, the code closes the socket and starts all over again, waiting for a new connection. As long as this code is running, you can connect and disconnect from your phone at will.

To test it out, you'll need an app that can transmit and receive data over Bluetooth. If you're on Android, download the BlueTerm+ application from the Google Play Store. After downloading, open the application and tap on the three dots in the lower-right corner to bring up the menu. Select Connect Device and choose your Intel Edison from the list of possible options. The application will connect, and the status in the upper-right corner will change. This is shown in Figure 4-1. Once connected, anything you type will be transmitted to your Edison and then sent back to you, one character at a time. Get it? It loops back.

Figure 4-1. *Example of using the Blueterm+ application before (left) and after (right) connecting to Intel Edison*

Open this file using your favorite text editor. Note that it's lines 45-47 where Intel Edison receives the characters from the Bluetooth transmission and loops them back out to the smartphone:

```
while True:
    data = server_sock.recv(1024)
    print("received: %s" % data)
server_sock.send("looping back: %s\n" % data)
```

Instead of looping sent items back, you're going to use the information to toggle the LED. I've modified the *SPP-loopback.py* for this purpose and put it on GitHub. Download it to your Intel Edison:

```
# wget https://raw.githubusercontent.com/smoyerman/Edison-
Bluetooth-LED/master/SPP-blink.py --no-check-certificate
```

Open this file in a text editor to observe the changes. Lines 18-22 import the mraa library and declare the LED pin as an output, just as you have in prior examples:

```
# Initiate LED stuff
import mraa
ledPin = 13        # LED pin number for mraa library
led = mraa.Gpio(ledPin)
led.dir(mraa.DIR_OUT)
```

Lines 51-59 replace the loopback functionality with LED controls:

```
data = server_sock.recv(1024)    ❶
if data == '1':    ❷
  led.write(1)
  stringy = "LED is on\n"
elif data == '0':    ❸
  led.write(0)
  stringy = "LED is off\n"
else:    ❹
  stringy = "Unrecognized Command. Please send either 0 or 1
to toggle the LED.\n"
server_sock.send(stringy)    ❺
```

❶ The data is first read from the socket as before.

❷ If the incoming data from the phone is a 1 (read by Python as the string `1`), Edison will turn the LED on, and the variable `stringy` will be updated accordingly.

❸ If the incoming data from the phone is a 0, Edison will turn the LED off, and the variable `stringy` will be updated accordingly.

❹ If any other data is received, the variable `stringy` will be set to an error message.

❺ The variable `stringy` is sent back to the phone as a status update.

Try it! Run the program as you did with the loopback file:

```
# python SPP-blink.py &
```

Then connect using the BlueTerm+ app and send the LED commands. You should see the onboard LED toggle on and off as you send 1s and 0s. You can build upon this very simple example by adding more commands to the *SPP-blink.py* script and programming them to take on different actions.

Killing Programs

If the old *SPP-loopback.py* is still running, and it will be if you haven't restarted your Edison, then you'll need to kill it before initiating the new one. To do so, you can use the `top` command, which provides a dynamic, real-time view of what's running on your system, and pipe the output to grep to find only what you're looking for. For example, to find where the `SPP-loopback.py` program is running, enter the following command:

```
# top | grep "SPP-loopback.py"
    378   225 root      S    16776   2%   0% python
SPP-loopback.py
```

Then kill the output with Ctrl+C. The first number in the returned line is the process ID of the `SPP-loopback.py` function. You can kill this function with the `kill` command:

```
# kill 378
```

If the program you're wanting to kill is a Python program, like *SPP_loopback.py*, then you can also kill it with the following:

```
# killall python
```

Be warned, however. If you have other Python programs running, this command will definitely kill those, too.

The Potentiometer

You can transfer the analog I/O code over to Python just as easily as you transferred the digital I/O code. First, assemble your Intel Edison and breadboard as shown in Figure 3-11. Next, create a Python script, *analog.py*, with the following contents:

```
import mraa as m        ❶

ledPinAnalog = 9        ❷
potPinAnalog = 5;   # Analog in potentiometer pin A5

pot = m.Aio(potPinAnalog)     ❸
led = m.Pwm(ledPinAnalog)
led.enable(True)

while(True):     ❹
    potReading = pot.readFloat()
    led.write(potReading)
```

❶ First, as always, you import the necessary mraa library.

❷ Just as in the Arduino example, you declare your pin numbers.

❸ Next, you initialize each pin with its respective function: analog input or PWM.

❹ Loop indefinitely, reading the potentiometer value at each iteration and assigning it to the LED. The only real difference between this code and the Arduino code is the lack of a `map` function. While Arduino expects an integer in the range 0 to 255 for PWM input, mraa expects a float in the range 0 to 1. By using the **`readFloat()`** function, you automatically get a 0 to 1 scaled decimal back from our potentiometer. For reference, the **`read()`** function will return the unscaled 0 to 1,023 integer value for analog in.

I2C Accelerometer

To recreate the accelerometer examples in Python, wire the MMA8451 to the Intel Edison breakout as shown in Figure 3-12. Just as you did in the Arduino example, the first step in reading data from the accelerometer is to perform a check that the device is connected and addressable. Create a Python script with the following code to perform that check:

```
import mraa as m

# Init I2C
x = m.I2c(1)       ❶
```

```
# Some addresses
MMA_i2caddr             = 0x1D      ❷
MMA8451_REG_WHOAMI      = 0x0D
MMA_DEVID               = 0x1A

# Check that we're connected to the MMA8451 chip
try:
    x.address(MMA_i2caddr)      ❸
    mma_id = x.readReg(MMA8451_REG_WHOAMI)      ❹
    if not mma_id == MMA_DEVID:      ❺
        print "Wrong device found! Dev ID = " + str(mma_id)
    else:      ❻
        "MMA8451 Detected!"
except:
    print "MMA Device Not Connected!"
```

❶ Create an I2C object using the mraa library.

❷ Declare variables for the addresses and registers of the accelerometer.

❸ Addresses the device at its known address.

❹ Ask the device for its ID.

❺ If the returned ID does not match the known ID of the MMA8451, then the script prints an error message indicating what was detected.

❻ If the returned ID matches the known ID of the MMA8451, the script prints the "MMA8451 Detected!" message.

Run the script, and you should receive the "MMA8541 Detected!" statement. If not, check your wiring.

Try Except

There is a new construction in your accelerometer code that has not yet been used in this chapter: `try` and `except`. The `try` command attempts to perform all the actions within its respective code block. If any of them throws an error, then the code exits the `try` block and moves to the `except` block. For this specific example, the code will throw an error if the MMA8541 is not addressable or cannot be read from. In that case, the code will print the error message, "MMA Device Not Connected!"

Installing the Dependencies

Now that you know you're connected to the accelerometer, it's time to read some data. To run this example, you'll need Numerical Python, or NumPy (*http://www.numpy.org/*). NumPy is really the base fundamental package for all scientific computing in Python. Install it with:

```
# opkg install python-numpy
```

Numerical Python is a great example of why you should do installs using opkg over pip at any available opportunity. Installing Numpy via pip will take several hours. Installing Numpy via opkg takes only several seconds.

Using the MMA Library

For this example, I've written a library and placed it on Github. Pull this library to your current directory, `cd` into the folder it creates, and `ls` to see the directory contents:

```
# git clone https://github.com/smoyerman/EdisonMMA8451.git
# cd EdisonMMA8451
# ls
MMA8451.py  README.md   example.py
```

Git Cloning

You may wonder why I used `git clone` this time whereas I've always used `wget` before. If there's more than one file in a directory, using `git` will grab all of the files in one fell swoop. By contrast, I generally use `wget` to pull one raw file at a time, avoiding readme and license files that I don't really need. It's a matter of preference really, but know that if there are multiple files in a directory or you're planning to update/resync these files later, `git` is probably the way to go.

The *example.py* file is the one you'll be using, and the *MMA8451.py* file is the library itself. Open *example.py* in a text editor, and you'll see that it's fairly similar to what you did in "I2C Accelerometer" on page 82 in your Arduino sketch. The contents are as follows:

```
import MMA8451 as mma8451      ❶
import time

# Potential range values      ❷
MMA8451_RANGE_8_G = 0b10      # +/- 8g
MMA8451_RANGE_4_G = 0b01      # +/- 4g
MMA8451_RANGE_2_G = 0b00      # +/- 2g (default value)

# Make accelerometer object
accel = mma8451.MMA8451()

# Check for MMA 8451      ❸
ismma = accel.check8451()
if ismma == True:
    print "MMA 8451 Found!"
else:
    print "No MMA Found. What is this?!"

# Set up mma, default range is 2Gs
accel.setup()         ❹
# Can declare a different range with
accel.setup(MMA8451_RANGE_8_G)

# Loop, read, and print out values
```

```
for i in range(10):     ❺
    ax,ay,az = accel.readData()
    print "(" + str(round(ax,1)) + ", "
              + str(round(ay,1)) + ", "
              + str(round(az,1)) + ")"
    orientation = accel.readOrientation()
    print orientation
    time.sleep(0.1)
```

❶ Import the mma library (it's within the mma library that NumPy is called).

❷ Sets variables for the different accelerometer ranges, and create the accelerometer object, accel.

❸ These six lines should seem vaguely familiar from the previous example. You're using the function check8451 to check that the identification register matches your expectations and then printing out a status message based on the function response.

❹ After checking the device, perform a setup function, which can take an optional input of accelerometer range. The code sets up once with the default of 2 Gs and then reconfigures to use 8 Gs instead.

❺ Read data values and device orientation from the accelerometer 10 times, printing the output of each to the screen.

When you run the code, you should see something like the following output:

```
# python example.py
MMA 8451 Found!
Range = 2Gs
Range = 8Gs
(4.4, 0.1, -4.2)
Portrait Up Front
(-0.0, 0.2, 1.0)
Portrait Up Front
(-0.0, 0.2, 1.0)
and more...
```

Just like before, if your MMA is lying flat on a table or other surface when you run this code, you'll see a value close to 1.0 for az, the third number in the parenthesis.

Accelerometers are frequently used for wearable devices such as step counters. The general idea is to log a portion of the data as it streams in and then to look for either a pattern or a certain event threshold (e.g., the force from the striking of a step). Intel Edison and the MMA8451 supply all the hardware you need to get started, and Python, with its easy access to libraries and data structures, provides the ideal environment to develop such an algorithm. Give it a shot! See if you can detect something a bit more obvious at first, such as tapping the MMA with your finger.

Troubleshooting

In playing with this previous example, I found that it was not uncommon for my MMA to stop working or not start up properly if I connected it when Intel Edison was already on and booted up. If this happens to you, shut down Edison, disconnect power from the board, and then reboot with the MMA already connected.

SPI Screen

Driving displays over Python has significant advantages over C or C++. Python has amazing image manipulation and image-drawing libraries that make configuring shapes and scenes a cinch.

As with the accelerometer example, I've written a library in Python for our display screen. Use `git` to grab it now:

```
# git clone https://github.com/smoyerman/EdisonILI9341.git
# cd EdisonILI9341
# ls
ILI9341.py          README.md          example_shapes.py
photo.py
```

Open the file *example_shapes.py* to see how drawing to the screen works:

```
""" Drawing something """

import ILI9341    ❶
```

```
disp = ILI9341.ILI9341()
disp.begin()    ❷

disp.clear((255, 0, 0))    ❸

drawing = disp.draw()    ❹

# Draw some shapes.    ❺
# Draw a blue ellipse with a green outline.
drawing.ellipse((10, 10, 110, 80), outline=(0,255,0),
fill=(0,0,255))

# Draw a purple rectangle with yellow outline.
drawing.rectangle((10, 90, 110, 160), outline=(255,255,0),
fill=(255,0,255))

# Draw a white X.
drawing.line((10, 170, 110, 230), fill=(255,255,255))
drawing.line((10, 230, 110, 170), fill=(255,255,255))

# Draw a cyan triangle with a black outline.
drawing.polygon([(10, 275), (110, 240), (110, 310)], out
line=(0,0,0), fill=(0,255,255))

disp.display()    ❻
```

❶ Import the ILI9341 library and create an ILI object called `disp`.

❷ Call the `begin` function, which sends the screen over 50 configuration commands and initializes the device.

❸ The function `clear` sets the screen buffer to the specified (R,G,B) values but does not yet send the buffer to the display. In the case of this script, the function is setting the background color to pure red. If no input is specified to `clear`, the function buffers a black background by default.

❹ The script gets interesting when you call the `draw` function, which returns a Python `ImageDraw` object. `ImageDraw` objects are essentially a canvas on which you can paint using `ImageDraw` functions.

The script then draws an ellipse, rectangle, two lines, and a polygon simply by specifying coordinates and colors of each shape.

❻ ❺ Finally, calling `display` renders the actual image to the screen.

A full list of `ImageDraw` functions and a couple of quick examples can be found on the documentation page (*http://effbot.org/imagingbook/imagedraw.htm*). You can even load images, such as JPEGs and PNGs, and draw on top of them. You'll do an example of this in Chapter 5.

Run the *example_shapes.py* file and, after a second or two of configuration and initilization, the screen will render the shapes given in the script. This library and script are meant as a starting point for building larger interactive systems. Python has libraries and distributions that handle building interactive games and images, so get making!

BLE Beacon

The final example in this chapter is turning your Intel Edison into a *Bluetooth Low Energy (BLE) beacon*, a transmitter that will broadcast signals that can be heard by compatible or smart devices. No additional hardware is necessary for this exercise, just an Intel Edison compute module and any breakout board. Because beacons are typically used for location sensing and tracking, this particular example works great when you put Intel Edison on a smaller breakout board. Imagine your Edison on the SparkFun base and battery blocks as a keychain and serving as a locater for your keys!

Edison Side

I've written a script to turn Edison into an ibeacon that is specific to Edison's Yocto distribution. It is based largely off of Donald Burr's awesome linux-ibeacon library (*https://github.com/dburr/linux-ibeacon*). Download this code from GitHub.

```
# wget https://raw.githubusercontent.com/smoyerman/edison-
ibeacon/master/ibeacon
```

You might notice that the *.py* extension is missing from this script. This is because the script is meant to be run as an executable. To run a Python script as an executable, you must specify the path to the Python interpreter installed on your platform as the first line of the script itself. In this case, the first line is as follows, since this is where Python resides on Edison:

```
#!/usr/bin/python
```

Note that this is the same thing you did for your shell script in "Scripting at Bootup" on page 105, except the extension there was */bin/sh*.

Next, set the permissions to run this file in executable mode using the `chmod` command:

```
# chmod +x ibeacon
```

Run the executable:

```
# ./ibeacon
Advertising on hci0 with:
        uuid: 0xE20A39F473F54BC4A12F17D1AD07A961
major/minor: 0/0 (0x0000/0x0000)
       power: 200 (0xC8)
LE set advertise enable on hci0 returned status 12
```

The script shows you the universally unique identifier (uuid) of your freshly created BLE beacon, as well as the major and minor identifiers and transmission power. In order to understand these parameters and use them optimally, it's best to explain what each one of them does:

UUID
: A 16-byte string used to label a large group of related beacons. For example, if I were a farmer and wanted to track each of my animals with an Edison beacon, I would assign them all the same uuid.

Major
: A 2-byte string used to distinguish smaller subset of the same uuid. On my farm, while all animals get the same uuid, pigs get one major identifier, and cows receive another.

Minor
: Another 2-byte string identifying individual beacons within the uuid and major sets. Though all pigs on my farm would

have the same uuid and major, each would receive its very own minor. How cute!

Power

The transmission (tx) power is used to determine the distance from the beacon. The specified tx power should be set to the strength of the signal exactly 1 meter from the device. This value must be calibrated and specified in advance so that a beacon can be used for a rough distance estimate.

You might now see why beacons are so cool and effective. Not only can I group and uniquely identify every single animal on my farm, I can also tell how far each one is from me (if I'm in range of the tx signal, of course). Additionally, BLE beacons work well indoors, unlike GPS and other satellite trackers.

Unblocking Bluetooth

If you run the beacon script and receive the following error, it's because your bluetooth is not enabled:

```
# ./ibeacon
Error: no such device: hci0 (try `hciconfig list')
```

Make sure to enable it before running the script:

```
# rfkill unblock bluetooth
```

You can set the default parameters in the ibeacon program at runtime. For a full list of the possible inputs, use the `help` flag:

```
# ./ibeacon --help
```

Smartphone Side

So, let's use Edison as a true BLE beacon! Download the Locate Beacon (or any comparable) app from either the App Store on iOS or the Google Play Store on Android. I like Locate Beacon because it's totally free and cross platform.

Open the app, and click the Locate iBeacons button. You'll need to tell Locate Beacon the specifics of your iBeacon in order for it to appear in the list. To do so, tap on the option that reads "Tap here to configure visible iBeacon UUIDs" and then on the plus sign in the upper-right corner of the next screen. The configuation screen will pop up for you to name your beacon (the name of your Edison works nicely here) and label it with the uuid, major, and minor identifiers. Note that the uuid you enter has to match the format of their example uuid exactly (include the "-" marks). Save your device and then go back one screen to the visible iBeacons. Your Edison should now appear in the list and show you a running distance to the device. Try moving around your room to see the distance change.

If you find that the distance is a bit off from what you'd expect (mine was pretty good with the default program settings), you can correct this on either the phone side or the Edison side. On the phone side, tap the device in the device list and then click the Calibrate button. The app will walk you through calibrating your distance. Alternatively, on the Edison side, you can run the program with different power settings until the phone readings line up with your real-life measurements. Setting the power lower will increase the distance read by the phone and vice versa.

Going Further

In this chapter, we've seen a lot of examples of the flexibility and power of Python programming. From general programming to algorithms to peripheral devices, Python's libraries and wide user community/support make it a really great environment for development and rapid prototyping. The following resources will help you to go further with Python and follow up on examples you've seen in this chapter:

Code Academy's introductory Python course (https://www.codecademy.com/tracks/python)
: I think the number of enrolled students more than speaks for itself: 2.5 million.

Python's own getting-started guide (https://www.python.org/about/gettingstarted/)
: Links to many resources and code examples for everyone from the newbie to the experienced Python programmer.

Intel Edison Products Bluetooth User Guide (http://bit.ly/edison-user)
: At 73 pages, this guide is not for the faint of heart! However, it is jam-packed with heaps of good information about Bluetooth hardware and software and Edison-specific knowledge.

5/Teach Edison to See

Vision is the art of seeing what is invisible to others. —Jonathan Swift

Introduction

One area where Intel Edison's compute power really shines is the area of *computer vision*: the gathering, processing, analyzing, and understanding of images and video data. Computer vision is a complex, widely studied, and ever-growing field in the scientific world today. The algorithms that comprise a computer vision system are often very computationally heavy tasks, making Intel Edison's dual-core x86 processor more aptly suited than other embedded systems. In fact, because Intel's x86 architecture is the top choice for high-performance computing (over 99% by market share), many available computer-vision algorithms have been optimized for this architecture and will perform better on an x86 than they will on other systems.

In this chapter, you'll be exploring some applications of computer vision on your Intel Edison using a live video stream as well as images pulled from the Internet. Along the way, you'll interface Edison with a webcam and learn to pull streaming video from that input.

Breakout Boards

While Chapter 3 and Chapter 4 were somewhat heavily dependent on using the Arduino Breakout Board, this chapter, along with Chapter 6, can be done almost in their entirety on either the SparkFun Base Block or the Mini Breakout Board. Since neither has a full-size USB port, you'll need a USB On-The-Go (OTG) cable that is microUSB male-to-full-sized-USB-female.

The SparkFun Base Block can be powered from either microUSB port, so you can power (and still connect to) it from the console port while using the other port for the OTG cable and device. However, the Intel Mini Breakout Board uses the same microUSB cable for power and OTG devices, so you'll need to use either the J21 power headers or solder on a barrel jack connector to supply power while using USB devices.

Materials List

Only two parts are required for this chapter:

1. A UVC-compatible webcam

   My personal preference is for the Creative Live! Cam Sync HD 720P Webcam, because it's small and quite cheap on Amazon (*http://bit.ly/camsync-720p*). However, Intel Edison has driver support for any UVC compatible camera. A complete list can be found at Ideas on Board (*http://www.ideasonboard.org/uvc/*).

2. A power supply

   If you're on the Arduino Breakout Board, USB power will deactivate when you select to use the full-size USB port for the camera. To power the board, you'll need to supply power to the barrel jack onboard instead. Any DC power supply in the range 7V to 15V will do; a good, fairly cheap example is Adafruit's 9V supply (*http://www.adafruit.com/products/*

63). Note that if you're on the SparkFun Base Block, you won't need to worry about a power supply because both microUSBs serve as power ports.

OpenCV

For much of the programming in this chapter, you'll be using OpenCV (*http://opencv.org/*) and its Python bindings. OpenCV stands for open source computer vision; it's open source because it's totally free and community developed. You can install openCV and its Python bindings with `opkg`:

```
# opkg install python-opencv
```

Extracting Colored Objects

The first example you'll do to get acquainted with OpenCV in Python is to detect the location of a colored object within an image. For this, use the interactive Python shell so that you can follow along with each command.

Open the interactive Python shell and then import the OpenCV library, numerical Python, and the urllib library (*https://docs.python.org/2/library/urllib.html*) for retrieving files from the Internet:

```
>>> import cv2
>>> import numpy as np
>>> import urllib
```

Use the urllib library to retrieve an image of a blue frog from Wikipedia (*http://bit.ly/edison-frog*) and save it in your current folder as *BlueFrog.jpg*. You'll notice that this is a high-quality photo; it will probably take a few seconds to download:

```
>>> urllib.urlretrieve('https://upload.wikimedia.org/wikipedia/
commons/6/6f/Dendrobates_azureus_%28Dendrobates_tinctorius
%29_Edit.jpg','BlueFrog.jpg')
```

This particular blue frog is the Dendrobates azureus, a poison dart frog from Brazil that gets its distinctive blue color from the poisonous alkaloids in its skin. You'll use this blue color to extract the skin of the frog from the rest of the image. First, use OpenCV to read the image from the file:

```
>>> frame = cv2.imread('BlueFrog.jpg')
```

Most images, such as JPEGs, are loaded as a 2D array of blue, green, and red (BGR) values in the range 0 to 255 that map the picture you see. To extract the blue frog, you need to tell openCV what range of colors you're specifically seeking. Define an array of the lowest BGR values that you would consider "blue" and an upper array of the highest values of "blue." Then, use the OpenCV `inRange` function to find all values within this range:

```
>>> lowerblue = np.array([50, 0, 0],dtype = "uint8")
>>> upperblue = np.array([255, 80,  80],dtype = "uint8")
>>> mask = cv2.inRange(frame, lowerblue, upperblue)
```

In this specific example, the value of blue must exceed 50, but the values of green and red cannot exceed 80. The `inRange` function then labels all the pixel locations within the image that meet this criteria. You can create an image showing only these pixels by applying the mask to your original blue frog with `bitwise_and`:

```
>>> res = cv2.bitwise_and(frame,frame, mask= mask)
```

Finally, to see your processed blue frog, you'll need to save the image. To save a single image of the original and processed photos next to each other, use numpy's `hstack` (horizontal stack) function while writing out the file:

```
>>>
cv2.imwrite('BlueFrog_processed.jpg',np.hstack((frame,res)))
```

Close the interactive Python shell.

Viewing Images

The only thing left is to view the processed image you've created. There are a few ways to do this.

Using the SPI Screen

Change directories into the `ILI9341` folder that we downloaded with `git` in Chapter 4. Within this folder is a script called `photo.py`. Open it in a text editor:

```
""" Showing a picture """
import Image      ❶
import ILI9341
```

```
disp = ILI9341.ILI9341()    ❷
disp.begin()

# Load the image
image = Image.open('/home/root/BlueFrog_processed.jpg')    ❸

# Resize the image and rotate it so it's 240x320 pixels.
image = image.rotate(90).resize((240, 320))    ❹

# Draw the image on the display hardware.
disp.display(image)    ❺
```

This file is the most basic file for displaying a photo on the screen. The script takes the following sequence of actions.

❶ It imports the necessary libraries: Python's imaging library and the SPI screen driver library.

❷ It creates the screen object and initializes it.

❸ It imports the image from the filesystem. Don't forget to change the path to this image if it differs from the default set here.

❹ It uses the image library to rotate and resize the image to fit on the screen.

❺ It displays the resized and rotated image to the screen.

Run this script, and you'll see the blue frog, before and after processing, appear on the screen.

This method is really simple and instantaneous, but it also has its drawbacks. Because the screen resolution is only 240 x 320 pixels, you'll either always be viewing your images using this maximum resolution or only seeing a small fraction of your images at a time. To see full-resolution images, anything over 240 x 320, you'll have to transfer the files to your host computer.

Using File Transfer

You can transfer the files back to your host computer using either of the two following methods:

Wireless transfer

You can `scp` or `sftp` the file back over to your host computer using the process described in "The Internet" on page 41.

MicroSD card

Plug a MicroSD card into the slot on Intel Edison, and move or copy the image over to the card:

```
# mv BlueFrog_processed.jpg /media/sdcard/
```

Then, before removing the device, unmount/eject it safely with the `umount` command:

```
# umount /media/sdcard/
```

Plug this card into your host computer to view the file.

The result is shown in Figure 5-1 for reference.

Figure 5-1. *The blue frog example image before (left) and after (right) processing*

Face Detection

Color detection is cool, but what about something more complicated? The next example detects human faces within an image. This is a tough processing problem for a number of reasons: faces differ in size, shape, color, and appearance and can appear at a variety of distances and angles within an image.

I've written a script for facial recognition that can be downloaded from GitHub:

```
# wget https://raw.githubusercontent.com/SCUMakerlab/Intel-
Edison-Code/master/FacialRecognition.py
```

After pulling the script, open it with your favorite text editor. The contents of the script are shown here:

```
# Import necessary libraries
import cv2      ❶
import urllib

# Grab the image
urllib.urlretrieve('http://stephaniemoyerman.com/wp-content/
uploads/2015/06/DSC_0713_linkedin.jpg', 'steph.jpg')     ❷

# And grab the config file for facial recognition
urllib.urlretrieve("https://raw.githubusercontent.com/Itseez/
opencv/master/data/haarcascades/haarcascade_frontal
face_alt.xml","haarcascade_frontalface_alt.xml")

# Use OpenCV to load the image
img = cv2.imread("steph.jpg")     ❸

# And convert to grayscale for processing
gray = cv2.cvtColor(img,cv2.COLOR_BGR2GRAY)

# Create the classifier and run the algorithm
faceCascade = cv2.CascadeClassifier("haarcascade_frontal
face_alt.xml")     ❹
faces = faceCascade.detectMultiScale(gray,scaleFactor=1.1,
        minNeighbors=5, minSize=(30, 30),
        flags = cv2.cv.CV_HAAR_SCALE_IMAGE)

# For each face found in the image, draw a box around it
for (x,y,w,h) in faces:     ❺
    cv2.rectangle(img,(x,y),(x+w,y+h),(255,0,0),2)

# Write the image back out to a file
cv2.imwrite("steph_facefound.png",img)     ❻
```

This script does a number of things:

❶ Imports the necessary libraries opencv and urllib.

❷ Utilizes urllib to download an image of me from my blog site and the opencv configuration file for face detection to the current directory. If you're curious about how complicated a face detection algorithm truly is, open the *haarcascade_frontalface_alt.xml* file and gaze upon its contents. The configuration for face detection is over a thousand lines long!

❸ Uses openCV to read in my image and then converts it to grayscale. This turns the image into a single valued array instead of keeping all three values of blue, green, and red. These sort of transforms are very common for image processing, and openCV has code for more than a hundred of them.

❹ Creates the `faceCascade` object using the parameters from the *haarcascade_frontalface_alt.xml* file. This object is used to find the location of my face using the `detectMultiScale` function. The parameters `scaleFactor`, `minNeighbors`, and `minSize` tell the algorithm the scaling between faces, the minimum distance between faces, and the minimum size of any single face respectively. Note that this algorithm is made to find as many faces as appear in an image; it's not limited to just one!

❺ The `for` loop draws a rectangular box around each face that was found in the image using the coordinates kicked out by the algorithm. These boxes are drawn on the original image and not on the grayscale version, though either is possible.

❻ The result is written out to *steph_facefound.png* in the current directory.

Run the Python script, and you'll notice that when it completes, three new files appear in your current directory: *steph.jpg*, *haarcascade_frontalface_alt.xml*, and *steph_facefound.png*. Use your favorite method to view to *steph_facefound.png* image. You should see me with a blue rectangle around my face as shown in Figure 5-2.

Figure 5-2. *Result of the OpenCV facial recognition example*

Using this example, you should now be able to perform face detection on any image you can pull from the Internet or load onto your Intel Edison. But, what if you want to do this with a live camera?

Webcam

The first step in live detection is to connect a live video feed into your Intel Edison. If you're on the Arduino breakout board, make sure to supply power via the barrel jack on the board, then flip the switch on the front of the board toward the full-size USB port and plug in your webcam.

Issue the `lsusb` command, which displays information about the USB buses on your system and the devices connected to them. You should see your camera appear in the list:

```
# lsusb
Bus 001 Device 002: ID 041e:4095 Creative Technology, Ltd
```

```
Bus 001 Device 001: ID 1d6b:0002 Linux Foundation 2.0 root hub
Bus 002 Device 001: ID 1d6b:0003 Linux Foundation 3.0 root hub
```

To check that the uvcvideo driver has successfully interfaced with your camera, use the `dmesg` (display or driver message) command, which prints the message buffer of the kernel. The `dmesg` command will output the entire buffer, which is a lot, so pipe the output to `grep`, searching specifically for the phrase uvc:

```
# dmesg | grep "uvc"
[   54.260633] uvcvideo: Found UVC 1.00 device Live! Cam Sync
HD VF0770 (041e:4095)
[   54.274395] usbcore: registered new interface driver
uvcvideo
```

Snapping Photos

Now that you're connected, you can use Python to grab control of your camera and snap a still image. Open the Python interactive shell and import OpenCV, which has excellent video-camera-binding functionality:

```
>>> import cv2
```

Connect to the video camera using the `VideoCapture` command. This command takes only one input, and it's the index of the video camera of interest. Since you presumably have only one camera connected, the index will be 0:

```
>>> cap = cv2.VideoCapture(0)
```

Multiple USB Devices

Although Edison only has one USB port, it can handle multiple connected devices through a powered USB hub. It's important that the hub have its own power supply, as Edison alone probably won't be able to supply enough current to power multiple USB devices.

You can grab a frame from the camera with the `read` function. You'll notice that the power LED on the camera comes on after issuing the `read` command. Be sure to hold up your camera and pose before doing this, because this is the selfie you're gonna get!

```
>>> ret, frame = cap.read()
```

To view the captured frame, you'll have to save it out to a file. Just as you did in previous examples, you'll do this using the function `imwrite`:

```
>>> cv2.imwrite('mypicture.png',frame)
```

Note that you can issue the `cap.read()` command as many times as you like before writing out the image. OpenCV will snap a frame every time you issue the command, and `imwrite` will write out the very last assignment of `frame`. Before closing the interactive Python shell, be sure to `release` your hold on the camera:

```
>>> cap.release()
>>> exit()
```

Check out your selfie by transferring the file over to your host computer or using your display screen.

Digital Camera

You may have noticed that we just positioned ourselves, took a photo on cue, and then displayed it out to an affixed screen. If you were to connect one of Edison's inputs to a button, you could program Python to do the same thing on a button press. Sound familiar? You've just created a digital camera!

Recording Video

You've just snapped a still photo, but you might be wondering how you can also record video from your camera. To do so, download my *VideoCapture.py* from GitHub:

```
# wget https://raw.githubusercontent.com/smoyerman/
EdisonOpenCVVideo/master/VideoCapture.py
```

Open the file in a text editor. You should see the following:

```
# Import Libraries
import numpy as np   ❶
```

```
import cv2

# Define variables    ❷
framerate = 20.0    # frames per second
videolength = 3     # length of video in seconds

# Grab Camera
cap = cv2.VideoCapture(0)    ❸

# Define the codec and create VideoWriter object
fourcc = cv2.cv.CV_FOURCC(*'XVID')    ❹
out = cv2.VideoWriter('myvideo.avi',fourcc,
       framerate, (640,480))

# Video part
for i in range(int(videolength*framerate)):    ❺
    if(cap.isOpened()):    ❻
        ret, frame = cap.read()    ❼
        if ret==True:    ❽
            frame = cv2.flip(frame,0)
            out.write(frame)
        else:
            continue

# Release the camera and video file
cap.release()    ❾
out.release()
```

This script performs the following sequence of events.

❶ Imports the necessary libraries: OpenCV and numpy.

❷ Defines the video recording variables: framerate and videolength.

❸ Binds to the camera.

❹ Creates an object for writing the video. The `VideoWriter` function takes four arguments: the output filename, the codec for video encryption (defined earlier in the code), the framerate, and the video width and height in pixels.

❺ Loops through the total number of frames required for the video.

❻ At each iterations of the loop, it checks that the camera is open and available.

7. If the camera is available, it reads a frame.
8. Checks that the frame was read properly. If it was, flip the frame and write it out to the video file. If it was not, simply skip this single iteration through the loop and `continue` with the next iteration.
9. Release both the camera and the video recording.

Point your camera at something interesting and run the script. The camera light should turn on shortly after the script initiates and off again about three seconds later when the script completes. Congratulations! You just shot your first video! Move the file to your host computer to check it out.

Video Codecs

Video codecs are a complicated beast. If you have problems playing your recorded video, try it in VLC Player. VLC Player has support for virtually every codec out there, is free, and runs on Mac, Linux, and Windows.

Streaming Video

This example uses OpenCV and Python's *Flask* library (*http://flask.pocoo.org/*) to create a video-streaming web server. First, install Flask using pip, since it's not available in the opkg repos:

```
# pip install flask
```

Next, I've written a short script that serves video content over the web. Download it from GitHub:

```
# wget https://raw.githubusercontent.com/SCUMakerlab/Intel-
Edison-Code/master/LiveStream.py
```

Open the script in a text editor to check out its contents:

```
from flask import Flask, Response     ❶
import cv2

class Camera(object):     ❷
    def __init__(self):     ❸
        self.cap = cv2.VideoCapture(0)
```

```
    def get_frame(self):     ❹
            ret, frame = self.cap.read()
            cv2.imwrite('blah.jpg',frame)
            return open('blah.jpg', 'rb').read()

app = Flask(__name__)    ❺

def gen(camera):    ❻
    while True:
        frame = camera.get_frame()
        yield (b'--frame\r\n'
               b'Content-Type: image/jpeg\r\n\r\n'
               + frame + b'\r\n')

@app.route('/')    ❼
def video_feed():    ❽
    return Response(gen(Camera()),
     mimetype='multipart/x-mixed-replace; boundary=frame')

if __name__ == '__main__':    ❾
    app.run(host='0.0.0.0', debug=True)
```

Amazingly, using the power of Python and its libraries, serving streaming video over the web reduces to only about 20 lines of code. Let's look at what it does.

❶ Imports the necessary libraries.

❷ Defines a `Camera` object that has two functions. The `class` call is used for defining an object in Python, whereas the `def` call is used for defining a function.

❸ Defines the function `init`, which happens the first time we call the `camera` class. It binds to the camera.

❹ Defines the function `get_frame`, which snaps a frame from the camera, saves it using `imwrite`, and then reads the saved file in a web-ready format.

❺ Initializes the webapp using Flask.

❻ Defines the `gen` function, which pulls the images from the camera and inserts them into content for a web page.

❼ Defines the URL path for this web page.

❽ Defines the function `video_feed`, which returns the web content.

❾ Initiates the app on your server in the main function and tells it to run at `0.0.0.0`, which is replaced at runtime with the IP address of your Intel Edison.

Close this file and run it. Since we specified `debug = True` in the `run` function, you'll see debugger messages in the Edison console. The output should match the following:

```
# python LiveStream.py
 * Running on http://0.0.0.0:5000/ (Press CTRL+C to quit)
 * Restarting with stat
```

On your host computer, head over to http://*XXX.XXX.X.XX*:5000 or http://*devicename*.local:5000 in your browser, where *XXX.XXX.X.XX* is your device IP address and ***devicename*** is whatever you named your Edison. You should see your video streaming live at this address. Mine is shown in Figure 5-3, along with me saying hi to you!

Figure 5-3. *A screenshot of live video from my web server*

You may notice that the video streams pretty well but that there's a little lag between the action and the video being displayed in your browser. This is due to the time it takes to save and reload the image in the script and the natural lag of the Internet information transfer.

Processed Streaming Video

As a final exercise, you'll create a web server that streams not only live video but also a processed version of that video. Both streams will be shown simultaneously for comparison. This is a great template for testing your own live-video processing scripts on Edison.

The processing that you'll be doing is known as edge detection. It sweeps across multiple directions and uses the rate of change of the colors in the image to determine when an edge has been hit. Writing such an algorithm from scratch requires a working knowledge of both calculus and matrix manipulations, but using OpenCV makes it a breeze.

First, you'll need to download a modified version of our last script from GitHub:

```
# wget https://raw.githubusercontent.com/smoyerman/
EdisonWebVideoProcessed/master/LiveStreamProcessed.py
```

Open the script in a text editor, and you'll notice that very little has changed; I've added lines 9, 10, and 15 and modified line 17. Above each change, I've also written in a comment for clarity. The following code snippet shows the top 18 lines of code from the file, where all the changes occur:

```
from flask import Flask, Response
import cv2
import numpy as np

class Camera(object):
    def __init__(self):
 self.cap = cv2.VideoCapture(0)
 # Reset camera capture size for faster processing   ❶
 self.cap.set(3,480)
 self.cap.set(4,360)

    def get_frame(self):
        ret, frame = self.cap.read()
  # Apply laplacian edge detection to image   ❷
```

```
        laplacian = cv2.Laplacian(frame,cv2.CV_64F)
    # Write out original and edge detected images at once    ❸
        cv2.imwrite('blah.jpg',np.hstack((frame,laplacian)))
        return open('blah.jpg', 'rb').read()
```

❶ Downsizes the image being pulled from the camera. This is just to keep the processing time to a minimum.

❷ Performs the entirety of our actual image processing. This one line applies the Laplacian filter for edge detection to our image.

❸ Using the same trick from the blue frog example, the code writes the before-and-after images out to the same file by stacking them next to each other. Then the web server loads both at once and displays them on the page.

Run the script and head over to the same browser location on your host computer as we did in the last example. You'll see a smaller image being served up by your camera on the left and the processed image served on the right. An example of my output is given in Figure 5-4.

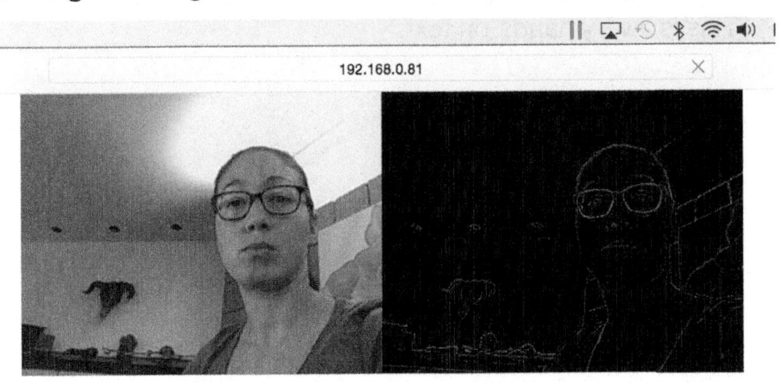

Figure 5-4. *A snapshot of my live web server displaying both the output of my webcam and the same image processed for edge detection*

Going Further

As you might be able to tell, image and video processing is a topic that is quite near and dear to my heart. If you'd like to explore these topics in more depth, there are a wealth of resources to get you started:

Python tutorials for OpenCV (http://bit.ly/opencv-py)
: This is a great place to start. There are many short, functional examples of image and video processing as well as good descriptions of the algorithms being employed. Be warned, however, that these tutorials run an older version of OpenCV than is currently available. Some of the examples will need small changes to work. Thank goodness for Google and Stack Overflow (*http://stackoverflow.com/*)!

Programming Computer Vision with Python (http://bit.ly/prog-computer-vision)
: A hands-on introduction to computer vision with Python. A lot of working examples and minimal deep-dives into the mathematical theory.

Learning OpenCV (http://bit.ly/learn-opencv)
: This text is great if you're looking to get way from just Python and explore the full OpenCV computing facilities. This is a very hands-on text.

6/Exploring Sound

From Wikipedia: hearing, auditory perception, or audition is the ability to perceive sound by detecting vibrations, changes in the pressure of the surrounding medium through time, through an organ such as the ear. Like touch, audition requires sensitivity to the movement of molecules in the world outside the organism. Both hearing and touch are types of mechanosensation.

Introduction

This chapter contains a lot of short examples and code snippets for doing various audio processing features, including installing audio programs, playing music files, recording and processing audio streams, performing speech recognition, and performing text-to-speech. A lot of these examples can be combined to build full end-to-end audio systems and voice-controlled electronics. A quick example of a voice controlled electronic system is given at the end of this chapter.

Materials List

There are only two parts required for this chapter: parts 1 and 2 below. As an alternative for part 1, you can use parts 3 and 4 in conjunction:

1. A Linux-compatible USB headset

I use the Logitech ClearChat Comfort/USB Headset H390. It's fairly cheap at $25 on Amazon and has a mute button that I find comes in handy. However, any Linux-compatible USB headset should work well with Edison, and cheaper ones are available.

2. A power supply

   Same requirements as those listed in Chapter 5 (see "Materials List" on page 126).

3. A USB-to-3.5 mm speaker/headphone and microphone jack

   If you prefer to use your (sure to be) already existent headphones, then you can buy USB to 3.5 mm jack converters. Good examples are the Plugable USB Audio Adapter with 3.5mm Speaker/Headphone and Microphone Jacks (*http://bit.ly/plugable-usb*) and the iLuv USB Audio Adapter (*http://bit.ly/iluv-adapter*).

4. A 3.5 mm microphone

   Assuming you already have 3.5 mm headphones, you'll also need a 3.5 mm jack microphone to plug into your USB adapter.

Connecting a Headset

The first step in exploring sound is being able to record sounds and listen to recorded sounds. Whereas in Chapter 5, you started video processing by pulling images from the Internet, here you'll start right away by connecting a headset.

Make sure the switch in between the microUSB and USB slot is flipped toward the USB port and plug in your USB headset. If you have the same (or similar) Logitech that I'm using, the red LED on the mute button will start blinking after a few seconds. This is an indication that the headset is receiving power from Edison and that the microphone is muted. If you push the button, the LED will turn solid red, indicating that the microphone is now active.

As a first step, you need to tell Edison to use your USB headset as the default system device for sound. You can tell how Edison references your sound card in the hardware with the aplay command:

```
# aplay -Ll | tail -5
  Subdevices: 1/1
  Subdevice #0: subdevice #0
card 2: Headset [Logitech USB Headset], device 0: USB Audio
[USB Audio]
  Subdevices: 1/1
  Subdevice #0: subdevice #0
```

Your device should appear as card 2 and will be called by name. The important piece is what comes after the colon and before the opening square bracket. This is the name by which you can reference your headset sound card; in this case, it's simply Head set.

Use your favorite text editor to open the */etc/asound.conf* file. Add the following line to the top, save it, and then close the file:

```
pcm.!default sysdefault:Headset
```

Remember to replace Headset with whatever device name you discovered using the aplay command.

Playing and Recording Sounds

Install the alsa utilities for sounds (*http://www.alsa-project.org*) and the text-to-speech engine (*http://espeak.sourceforge.net/*) espeak:

```
# opkg install alsa-utils espeak
```

OPKG Installing

Occasionally you'll get errors using the opkg installer, most of which can be solved by updating opkg:

```
# opkg update
```

Alsa automatically installs some wave files for testing your headset. You can play one of these to make sure your audio is configured properly:

```
# aplay /usr/share/sounds/alsa/Front_Center.wav
```

If everything is installed properly, you should hear the words "Front, Center" in a nice-sounding woman's voice. Let's switch that to a creepy robot voice by using espeak!

```
# espeak "Front, Center"
```

If you just type the command espeak, you can issue creepy robot text words until you terminate with Ctrl+C:

```
# espeak
front
center
hello
whatever
^C
#
```

The espeak program has a lot of fun options to play around with, from intonation to speech speed and even language! You can see the full list by calling the help dialogue (the same flag works for aplay and arecord and many other command-line programs):

```
# espeak -h
```

To give you a feel for it, try asking, "How are you?" in fast-paced, high-pitched Swedish:

```
# espeak -v sw "Hur mor du?" -p 99 -s 250
```

You can also record your own voice using the arecord command and then play it back using aplay. Without any specified arguments, arecord will record forever. You can kill it with Ctrl+C:

```
# arecord hello.wav
Recording WAVE 'hello.wav' : Unsigned 8 bit, Rate 8000 Hz, Mono
^CAborted by signal Interrupt...
# aplay hello.wav
Playing WAVE 'hello.wav' : Unsigned 8 bit, Rate 8000 Hz, Mono
```

Makeshift MP3 Player

Remember how awesome the iPod was when it came out? You can turn Edison into your very own iPod quite easily.

Form Factor

If you'd like to turn Edison into a small form-factor mp3 player, one great setup option is the SparkFun Base and Battery Blocks. From here, you can tack on the OLED Block for display controls or the microSD Block for increased storage capacity. If you'd like your music player to last for a really long time, a cute trick is to connect a cell phone battery pack (*http://bit.ly/ex-battery*) to one of the microUSBs instead of using the Battery Block. These have a much higher battery capacity.

First, install the `mpg123` library (*http://www.mpg123.de/*) for playing mp3s:

```
# opkg install mpg123
```

Next, just for this example, you'll pull some songs from LastFM's free download collection:

```
# wget http://freedownloads.last.fm/download/59565166/From
%2BEmbrace%2BTo%2BEmbrace.mp3
# wget http://freedownloads.last.fm/download/569330114/Lost
%2BBoys.mp3
# wget http://freedownloads.last.fm/download/569264057/Get
%2BGot.mp3
```

The downloaded filenames are shown here:

```
#  ls *.mp3
From+Embrace+To+Embrace.mp3  Lost+Boys.mp3
Get+Got.mp3
```

You can play any one of these files with the `mpg123` command and quit with Ctrl+C:

```
# mpg123 Get+Got.mp3
```

You can play all the mp3 files on shuffle with the `-Z` flag and the wildcard *:

```
# mpg123 -Z *.mp3
```

You can create playlist for these files using a simple text document. Create a file called *playlist.txt* with the following contents.

Remember to replace */home/root/* with the complete paths to the files on your own system:

```
/home/root/Get+Got.mp3
/home/root/From+Embrace+To+Embrace.mp3
/home/root/Lost+Boys.mp3
```

Now play your playlist with mpg123:

```
# mpg123 -@ playlist.txt
```

You've just created an iPod! You can even use your SPI screen to display the info for each song.

Recording Audio with Python

Let's do something a little more computationally savvy with our audio. You'll use Python to interface with the audio stream and detect when you're speaking into the mic and when you're silent.

You'll use the pyaudio library (*https://people.csail.mit.edu/hubert/pyaudio/*) for this. First, install the dependencies using opkg:

```
# opkg install libjack
# opkg install --nodeps jack-dev
# opkg install libportaudio-dev
```

Then, install pyaudio, making sure to set the flags so that pip trusts the unverified pyaudio library:

```
# pip install --allow-external pyaudio --allow-unverified
pyaudio pyaudio
```

Basic Recording

Now, download a simple audio-recording file from GitHub, compliments of Mahmoud Abdrabo. This is a modified example from the *pyaudio* documentation (*http://bit.ly/pyaudio-docs*):

```
# wget https://gist.githubusercontent.com/mabdrabo/8678538/raw/
30e63a8c2ab78b516b13a180895308b8a4244ecf/sound_recorder.py
```

This will download sound_recorder.py. Open this file.

```
import pyaudio    ❶
import wave
```

```
FORMAT = pyaudio.paInt16     ❷
CHANNELS = 2
RATE = 44100
CHUNK = 1024
RECORD_SECONDS = 5
WAVE_OUTPUT_FILENAME = "file.wav"

audio = pyaudio.PyAudio()     ❸

# start Recording
stream = audio.open(format=FORMAT, channels=CHANNELS,
                rate=RATE, input=True,
                frames_per_buffer=CHUNK)
print "recording..."
frames = []

for i in range(0, int(RATE / CHUNK * RECORD_SECONDS)):     ❹
    data = stream.read(CHUNK)
    frames.append(data)
print "finished recording"


# stop Recording
stream.stop_stream()     ❺
stream.close()
audio.terminate()

waveFile = wave.open(WAVE_OUTPUT_FILENAME, 'wb')     ❻
waveFile.setnchannels(CHANNELS)
waveFile.setsampwidth(audio.get_sample_size(FORMAT))
waveFile.setframerate(RATE)
waveFile.writeframes(b''.join(frames))
waveFile.close()
```

❶ The script first imports the necessary libraries: `pyaudio` to pull the audio stream, and `wave` to save the audio file.

❷ Next, it sets the audio recording format by declaring variables for the log format (`FORMAT`), data rate (`RATE`), amount of data to read at a time (`CHUNK`), duration of recording (`RECORD_SECONDS`), and an output filename (`WAVE_OUTPUT_FILENAME`).

❸ The script then creates a `pyaudio` audio stream and opens it with the given configuration. Before the script starts recording, it prints "recording..." and then creates an empty frame array.

❹ Each time through the for loop, which iterates as many times as there are chunks to read in five seconds, the audio data that is read is tacked onto (or appended to) the frames array. When the loop finishes, "finished recording" is printed.

❺ The audio stream is stopped and closed, and the pyaudio object is terminated.

❻ The full five-second recorded audio is written to disk using wave functions. The file is opened and configured, the frames array is written out, and the file is closed.

Run the file and, making sure your mic is not muted, record yourself speaking in between the "recording..." and "finished recording" lines. Play your wonderful recording back to you using aplay:

```
# aplay file.wav
```

Thresholding

It's great to just blanket record sounds, but it would be nice to make our system a bit smarter. We're going to modify sound_recorder.py to help us do that.

First, add the following line at the end of the import statements:

```
import audioop
```

Now, in the for loop where you recorded the stream, add the print statement in between the the two other lines as shown here:

```
data = stream.read(CHUNK)
print audioop.rms(data,2)
frames.append(data)
```

The audioop library (*https://docs.python.org/2/library/audioop.html*) contains a series of functions and classes to help you manipulate raw audio data. In this case, you're taking the *root mean square* (RMS) of each audio chunk you read. The root mean square is calculated by squaring every data point in the sample, taking the mean of these values, and then taking the square root of that mean. You can think of it like the mean, except the RMS value puts a higher emphasis on outlier points.

The RMS value is a common way to detect speech and silence in audio streams. If even just a few points have higher volumes, the RMS value will be skewed way up. Save your modified `sound_recorder.py` file and run it again. This time, as the data is processed, the RMS values will be printed to the screen.

Run the script again with your mic on mute. You'll notice that the printed values drop immediately down to low single-digit values. The audio stream is receiving basically 0 noise. Run it again with the mic unmuted but without speaking. You'll probably see values in the 100s and 200s (if your house is approximately as noisy as mine). If you put the headset on and begin talking, you'll notice that the values probably shoot up into the 1,000s. If you're using a USB headset, one of the reasons this value shoots up so high is that your mic is directional, meaning it prefers sound to come from one direction as opposed to all over. In this case, it's from the slot facing your mouth.

By choosing an adequate threshold, you can actually detect fairly well when people are speaking and when they're silent. Try playing with this yourself. Open the `sound_recorder.py` script one more time, define a threshold for speech, and modify the loop as follows:

```
threshold = 800
for i in range(0, int(RATE / CHUNK * RECORD_SECONDS)):
    data = stream.read(CHUNK)
    rms = audioop.rms(data,2)
    if rms > threshold:
        print "You're speaking"
    frames.append(data)
```

Save and exit the file, and then run it. This time through, silence gets you nothing, but any time you speak, you should see the "You're speaking" text. Feel free to choose a threshold that best suits your headset; it will definitely vary by model and even by fit. If you find yourself needing more time, modify the recording duration (`RECORD_SECONDS` variable) to better suit your needs.

Speech Recognition

If you think speech detection is cool, you're in for a treat, because speech recognition blows it out of the water. Up until now, we've been doing everything offline, directly on Edison without the need for Internet. This example uses the Google Speech Recognition API and so needs an Internet connection to work. Make sure your Edison is on WiFi.

The Python library you'll be using is SpeechRecognition (*https://pypi.python.org/pypi/SpeechRecognition/*). You'll need to install a single dependency and `pip install` the library itself:

```
# opkg install flac-dev
# pip install SpeechRecognition
```

After installation, create a Python script called `SpeechReconize.py` with the following contents:

This code comes almost directly from the examples on the SpeechRecognition page (*https://pypi.python.org/pypi/SpeechRecognition/*).

```
import speech_recognition as sr

r = sr.Recognizer()
print "Start Listening..."
with sr.Microphone() as source:
# use system default microphone as the
# audio source

    audio = r.listen(source)
    # listen for the first phrase and extract it
    # into audio data

print "Done Listening..."
try:
    print("You said: " + r.recognize(audio))
    # recognize speech using
    # Google Speech Recognition

except LookupError:
```

```
        # speech is unintelligible
        print("Could not understand audio")
```

Run the script now. You'll see that the script hangs after it prints "Start Listening..." until you both start and finish speaking. When you finish, the script prints "Done Listening...", hangs for another second or so, and then prints out the contents of your speech or the error message, "Could not understand audio." If you play with this script a little bit, you'll immediately see how powerful it is. The `listen` command is very good at capturing full phrases, and the speech classifier is very, very accurate. Essentially, you've just added Google Now to your Edison in less than 20 lines of code.

It's worth noting that this example uses `pyAudio` to handle the audio stream from the mic. The thresholding and audio stream capture is done in much in the same way as you did manually in the previous example, but this time you've got a fancy library to do it all in the background!

This does raise an interesting point though. In the last exercise, I told you that ambient noise and speech volume would vary with not only the environment but also with the microphone. You might wonder how the `SpeechRecognition` library works so well just straight off the bat. `SpeechRecognition` has some default settings that work pretty well for most normal situations. Having a nice, directional headset microphone makes the algorithm work even better. Think about how much more defined your voice is into this headset than it would be yelling at your phone on a New York street corner. However, if you are out on that New York street corner, you're going to need a way to recalibrate as we did prevoiusly. All you need to do is add a single line to the script, directly above the `listen` command:

```
with sr.Microphone() as source:
     # use system default microphone
     # as the audio source
    r.adjust_for_ambient_noise(source)
     # adjust thresholding
    audio = r.listen(source)
     # listen for the first
phrase                                              # and
extract it into audio data
```

This adjust_for_ambient_noise function listens for one second to the audio stream and uses that to calibrate the threshold for ambient noise. Now, in theory, your recognition should be good in almost any environment.

Controlling Devices

There's one last example in this book, and it's using speech recognition to control peripheral devices, in this case your SPI screen. Connect the SPI screen to the Arduino breakout board now. You'll use the SpeechRecognition library for parsing speech and the ILI9341 library to drive the SPI screen. The script will take in your speech, parse it, and then use it to drive the color of the screen.

I've posted a quick example of this system to GitHub. In order for this example to work as is, you have to download it into the *EdisonILI9341* directory or wherever your ILI9341.py file resides if you moved it. Change into that directory now, then pull the script using wget:

```
# wget https://raw.githubusercontent.com/smoyerman/
VoiceControlledScreen/master/ScreenControl.py
```

Open this file in a text editor. It should look fairly similar to the previous example, but expanded:

```
import speech_recognition as sr     ❶
import ILI9341

# Construct screen and speech recognizer
disp = ILI9341.ILI9341()     ❷
disp.begin()
r = sr.Recognizer()

# Hard-coded colors
red = (255,0,0)     ❸
green = (0,255,0)
blue = (0,0,255)
white = (255,255,255)
black = (0,0,0)
puple = (100,0,100)

# Loop and listen 10 times
for i in range(10):     ❹
```

```
    # Listen for audio each time
    print "Start Listening..."
    with sr.Microphone() as source:
        # Only check for ambient noise the first time
        if i == 0:
            r.adjust_for_ambient_noise(source)
        audio = r.listen(source)
    print "Done Listening..."

    try:
        # Parse text
        speechString = r.recognize(audio)    ❺
        print("You said " + speechString)
        speechArray = speechString.split()    ❻
        # Check for colors
        if "red" in speechArray: color = red
        elif "blue" in speechArray: color = blue
        elif "green" in speechArray: color = green
        elif "purple" in speechArray: color = purple
        elif "white" in speechArray: color = white
        elif "black" in speechArray: color = black
        # Display to screen
        disp.clear(color)    ❼
        disp.display()

    # Check for error
    except LookupError:
        print("Could not understand audio")
```

There are several new components to this script:

❶ At the top, the script imports ILI9341.

❷ It declares the screen object and initializes the screen.

❸ After initializing the speech recognition object, colors are hardcoded as different RGB tuples.

❹ The speech recognition is then placed in a loop that will run 10 times. Each time, we process audio the same way as in the previous example.

❺ When the processing is finished, we store the output as the variable `speechString`.

❻ The `speechString` variable is then split into an array, so "An expression like this" becomes ["An","expression", "like","this"]. This is to keep words like "bluebird" from triggering the phrase blue. By splitting it into an array, the script is forcing an exact match. The if and elif statements then perform the color keyword searching.

7. Finally, disp.clear(color) sets the color, and disp.display() renders it to the screen.

Run the code, and you'll see the screen change with your voice commands:

```
# python ScreenControl.py
Start Listening...
Done Listening...
You said I see a red door  <-- screen turns red
Start Listening...
Done Listening...
You said and I want to paint it black  <-- screen turns black
Start Listening...
```

if and elif

The way the code is written is like going down a queue. The code will only check for the next keyword if all other conditions before it have not been met. This is the nature of elif. Therefore, the expressions "red is black" and "black is red" will both do the same thing: render the screen red.

Going Further

There are two main areas of study for going further in this chapter: digital signal processing, which encompasses audio signals, and natural language processing, which covers topics like automatic speech recognition and speech to text:

Digital Signal Processing (https://en.wikibooks.org/wiki/Digital_Signal_Processing)
: A free wikibook starting with the basics of digital signals and moving up into some very complicated territory. It contains a lot of hands-on exercises in Matlab and its free counterpart, Octave. Many other free online resources such as this

one exist; do a quick Google search to find which works best for you.

Natural Language Processing with Python (http://bit.ly/natural-lang)
A hands-on text with a lot of examples. Note that this book can get a little heavy on theory and upper-level mathematics.

7/Conclusions

It's closing time; you don't have to go home, but you can't stay here.

This book is meant to be an introduction to the wide world of possibilities using Intel Edison: Linux, programming, sensors, Bluetooth, WiFi, and especially combinations of them. When all of these ideas can be fused together in one system, that's when interesting new concepts and ideas can become reality. Intel Edison is the first true maker device that provides this possibility right out of the box.

This book is not, by any means, a comprehensive overview of every aspect of Intel Edison. The WiFi and Bluetooth alone would probably fill the book, and more text books have been written about image and audio processing than I care to count. But if you're looking to build a cool end-to-end system incorporating all of these elements (and avoid picking up a degree in electrical engineering), this book will definitely get you started.

However, there are a few topics about which I should definitely say a few more words.

Linux Flavors

The official Intel Edison operating system, as supported by Intel, is Yocto Linux. While Yocto is a powerful operating system for embedded devices, some people find it easier to work with a fuller and more widely used Linux distribution. For this reason, the Intel Edison community worked hard to bring up Debian (aka Ubilinux) on Edison. Sparkfun has written a quite comprehensive tutorial (*http://bit.ly/ubilinux*) on installing Ubilinux on Intel Edison. Note that what the tutorial says is absolutely true: you can definitely brick your Edison if you don't follow the instructions properly. So, be warned and be diligent if you plan on installing Debian.

There are two real advantages of Debian aside from the fuller feature set of the OS itself: the much wider user community and the different package manager. A quick Google search will show you that Debian has many more users than Yocto, meaning problems will be easier to track down and debug when you run into them. But, in my opinion, the real reason to switch from Yocto to Ubilinux is the package manager.

In its most common use, Yocto Linux is built from scratch with the minimal set of dependencies necessary for a given project. This lightweight build is then flashed onto the device for use. While this keeps the distribution extremely lightweight and configurable, it necessitates a rebuild of the operating system every time a new package must be added. The rebuilds and reflashes reset your configuration changes back to the Yocto defaults, just like reflashing your Edison does. While this model works for production—you use one test system to figure out exactly what you need and then use that as the base image for all devices—it's not exactly ideal in the maker space. As makers, we want more options and definitely less time-consuming ones.

That's not it for Yocto, though. As you've seen in this book, Yocto does have the option of using `opkg` to manage packages instead of performing builds and rebuilds. Simply edit your configuration files to tell `opkg` where to find the necessary libraries and then you can pull and install them as needed. But did you ever stop to wonder where all those libraries came from? The answer is simple: they come from a single Intel employee (and a god among men), Alex T. Alex recognized that makers aren't going to want to build and rebuild the system from scratch every time they need a new package. Instead, he built a *huge* set of Yocto Linux packages for Intel Edison and made them freely available online. The links you added to your *base-feeds.conf* file are links to his prebuilt packages. He's the reason we were so easily able to perform all the software tasks we have in this book. Thanks Alex! I tip my hat to you.

But this does beg the question, what if Alex T. hadn't come along? How would we have installed packages then? Do all distributions work this way, with one person holding the reigns? Obviously, the answer is no. Linux *can't* work that way; it would just be too difficult. Most Linux distributions have an officially released package manager that comes prelinked to a very full and well-maintained library specific to that OS. Debian uses Ubuntu's Advanced Packaging Tool (APT) to manage package installs and the `apt-get` utility for command-line calls. To say that this utility is well supported is an extreme undersell. Currently, Debian has over 43,000 supported packages. And by installing Ubilinux on your Edison, you have immediate access to all of them, for free, forever. It might be worth switching to Ubilinux if you begin a project on Yocto that requires packages that you're not able to or are incredibly difficult to install on Yocto.

Programming Languages

Obviously, this book is quite Python-heavy for all the reasons mentioned in Chapter 4 (see "Introduction" on page 97) and also because of my own personal preference for Python. But there are other options worth exploring on Edison.

Node.js

Node is a more recent Java-based language that has accumulated a huge following in a very short period of time. According to the Node.js website (*https://nodejs.org/*), "Node.js is a platform built on Chrome's JavaScript runtime for easily building fast, scalable network applications. Node.js uses an event-driven, non-blocking I/O model that makes it lightweight and efficient, perfect for data-intensive real-time applications that run across distributed devices."

In my opinion, Node.js is a little bit more difficult to wrap your head around than Python. The syntax of Node.js is different than a lot of other coding languages (especially Python), and the nonblocking element definitely takes a bit of getting used to. Whereas Python (and most other programming languages) moves linearly through your code, waiting until one command set is complete before executing the next (i.e., blocking), Node runs asynchronously and can execute tasks in a nonlinear fashion (i.e., nonblocking). Although this is a bit odd to think about, it's an extremely useful architecture when dealing with distributed systems across many devices. While you're waiting for interdevice communication, it's very useful if the waiting doesn't block all the other processes in the code. A great example of these sorts of distrbuted systems are web servers and real-time web applications.

If you're planning to use your Edison to host a web service, it's definitely worth looking into Node.js to see if it meets your needs. While there are also Python packages that can do the job, such as Flask and Tornado (which is actually asynchronous and very similar to Node.js), Node is preinstalled on Edison and might naturally be the best fit for the job. If you'd like to learn Node.js, the best way is to use Node to install a tutorial library and actively work though the examples.

The Node.js package comes with a preinstalled package manager (like Python's pip), *npm*. To install the tutorial library on your Edison, simply run:

```
# npm install -g learnyounode
```

And then run the exercises in the tutorial with:

```
# learnyounode
```

The preceding example is the "Hello, World!" of Node tutorials, but many more can be found at the NodeSchool website (*http://nodeschool.io/*). They cover everything from JavaScript to asyncronous I/O to using `git`.

C and C++

Through your Arduino programming in Chapter 3, you experienced a little bit of what C is all about. While you can program C++ code into the Arduino IDE, you can also write standalone C++ code in Linux, compile it, and then run it all natively as well. This is one of the reasons that Intel Edison is such an amazing learning tool for programming.

Some Intel-Edison-specific hardware is already supported in this fashion. For example, the SparkFun OLED block for Edison (*http://bit.ly/oled-block*) has only a standalone C++ library and no Arduino-style libraries to speak of. And although we've been accessing Intel's mraa library in Python, it's actually natively written in C, and we're just calling it through Python bindings.

C is the required programming language for most embedded devices, and Intel Edison presents a great opportunity to learn how to use C on an embedded processor without being constrained by the memory or power of the usual embedded systems. When you get into C, you'll notice that it's *super fast*. Compiled languages just tend to be that way, making them advantageous if you're looking for computational speed or timed actions.

While writing an introduction to C is definitely beyond the scope of a single section of a single chapter of this book, I can leave you with examples of `HelloWorld` in both C and C++, along with some compilation help. You'll see immediately why I love Python so much.

For C, create a file named *HelloWorld.c*. Paste in the following contents:

```
/* Hello World program */
#include<stdio.h>

main()
{
    printf("Hello World\n");
}
```

Every program in C must contain a `main` function that is executed when the program is run. Before running, however, this code must be compiled. On Edison, you can compile C code with the following command:

```
# gcc InputFile.c -o OutputFileName
```

In this example, to compile your `HelloWorld.c` code into an executable program called `HelloWorld`, you would issue the following:

```
# gcc HelloWorld.c -o HelloWorld
```

Then, run the compiled program:

```
# ./HelloWorld
Hello World
```

The C++ code follows the same general format. It must also contain a *main* function, except the *main* function must return an integer and be declared as such. The standard *stdio.h* library used in the C example will still work in this code, but C++ has a slightly nicer input/output library that should be used instead.

```
// 'Hello World!' program
#include <iostream>

int main()
{
  std::cout << "Hello World!" << std::endl;
  return 0;
}
```

Compile this code and run it. You'll notice that the compilation call is almost exactly the same, except the compiler itself is now g++ instead of gcc:

```
# g++ HelloWorld.cpp -o HelloWorldCPP
# ./HelloWorldCPP
Hello World!
```

If that was fun for you, the following tutorials can take you a whole lot further:

- C (*http://www.cprogramming.com/tutorial/c-tutorial.html*)
- C++ (*http://www.cprogramming.com/tutorial/c++-tutorial.html*)

The Intel XDK IoT Edition

Intel has developed an XDK for Galileo and Edison that wasn't really discussed in this book. If you're interested in playing with it, you can download and install it from the Intel website (*http://bit.ly/intel-xdk*).

The main idea behind the XDK is that instead of coding directly, you can use a graphical user interface to select sensors and basic system functionality and the XDK will generate the code for you. The main reason I don't discuss the XDK in this book is because of my own belief that learning should take you deeper into how things work. Getting into Linux, Python, circuits, and Arduino helps to do that. The XDK, on the other hand, limits you to a more shallow layer. It hides the code and the inner workings of the device behind a GUI. It makes it hard *to learn*.

That said, it doesn't make it hard to prototype. Don't be put off from playing with the XDK because it's simple. Simple can be good. Simple can be very fast.

Shutdown Now

As you can see, there is so much more to Edison than meets the eye. It's not an Arduino or a Galileo or a Raspberry Pi—this little module is something entirely new. So let your Edison take on its own personality and inspire your new and creative projects. Happy making!

A/Materials

The following is a list of materials that you'll need to complete all of the exercises in this book in their entirety. If you buy them all, it will cost you approximately $200, depending on which brands you opt for and where you buy them:

Edison compute module and Arduino Breakout Board
: For this book, we focus on Intel Edison and the Edison Arduino Breakout Board. You can buy the set for approximately $100 USD at any of the following locations: Maker Shed (*http://www.makershed.com/products/intel-edison-kit-for-arduino*), SparkFun (*https://www.sparkfun.com/products/13097*), Mouser Electronics (*http://www.mouser.com/new/Intel/intel-edison/*), Adafruit (*http://www.adafruit.com/product/2180*), Seeedstudio (*http://www.seeedstudio.com/depot/Intel-Edison-for-Arduino-p-2149.html*), and Amazon (*http://amzn.com/B00ND1KH42*).

Two microUSB cables
: Make sure to buy high-quality microUSB cables for powering the board; substandard ones (like the ones you often get for charging cell phones and other electronics) probably won't work. MicroUSB cables can be found easily on Amazon or many other sites if you don't have them already.

DC power supply
: Any supply between 7V and 15V with 500 mA or more will do. You can find fairly cheap ones on Amazon (*http://bit.ly/1a-pa*) that will work.

Breadboard
: A half-size breadboard with power rails is probably the most well suited to this book. These can be found for a few dollars on Amazon (*http://bit.ly/BB400-board*). You can also purchase them from Adafruit (*http://www.adafruit.com/product/64*), but they're slightly more expensive at $5.

Male-to-male breadboard wires

These are probably cheapest on Amazon (*http://bit.ly/j-wires-mm*). Adafruit also sells packs in two different flavors: regular in mixed lengths (*http://www.adafruit.com/product/153*) and premium in just one length (*https://www.adafruit.com/products/760*). They range in price from $6 to $8.

An LED

You'll need to buy a pack because LEDs are rarely sold as individual units. Adafruit (*http://www.adafruit.com/categories/90*) has heaps of them, and you're looking specifically for a bare breadboard LED in a single color (*http://www.adafruit.com/products/299*). You can also buy mixed packs on Amazon (*http://bit.ly/2pin-led*).

Resistors

A pack of resistors (*http://bit.ly/haobase-1k*) with any value from 220 ohms to 1 kohms will work. However, if you're thinking about getting more seriously into building and making, try getting a mixed pack in a large range, such as those on Amazon (*http://bit.ly/16v-resistor*).

Button

The cheapest and easiest are sold as a pack on Adafruit (*https://www.adafruit.com/products/367*) for $2.50. However, button choices are fairly limitless, and you can buy some really cool ones (*http://www.adafruit.com/category/235*) if you're so inclined.

Potentiometer

You can buy a single potentiometer from Adafruit (*https://www.adafruit.com/products/356*) for $1.25.

Accelerometer

The specific accelerometer chip you'll be using in this book is the MMA8451 sold for $7.95 on Adafruit (*http://www.adafruit.com/products/2019*). Accelerometers are one of the most widely used and basic sensors for embedded devices, which I why I chose them to pair with Edison.

SPI-driven display
: Any screen that uses ILI9341 will work with the exercises in this book. I personally recommend the resistive touch shield (*http://www.adafruit.com/products/1651*) (more on why in "SPI Screen" on page 87) or the resistive touch breakout (*http://www.adafruit.com/products/1770*).

UVC-compatible webcam
: My personal preference is for the Creative Live! Cam Sync HD 720P Webcam, because it's small and quite cheap on Amazon (*http://bit.ly/camsync-720p*). However, Intel Edison has driver support for any UVC-compatible camera. A complete list can be found here (*http://www.ideasonboard.org/uvc/*).

Linux-compatible USB headset
: I use the Logitech ClearChat Comfort/USB Headset H390. It's fairly cheap at $25 on Amazon and has a mute button, which I find comes in handy. However, any Linux-compatible USB headset should work well with Edison, and cheaper ones are available.

Glossary

analog-to-digital converter (ADC)
A device that takes a continuous analog signal and discretizes it into a digital signal.

anode
The lower voltage side of a polarized electrical device.

Arduino
A paired hardware and software platform that enables rapid design, prototyping, and building of devices that can sense and control aspects of the physical world.

Arduino breakout board
An Arduino-compatible breakout for Edison. The pinouts on this breakout match the standard Arduino Uno configuration. The Intel Arduino Breakout Board, SparkFun Base Block, and Intel Mini Breakout Board are the three main breakout boards for Intel Edison. See also Intel Mini Breakout, SparkFun Base Block.

avrlibc
A C library to use with GNU Compiler Collection (GCC) on Atmel AVR microcontrollers. See also Arduino.

baud rate
The rate at which connected devices perform serial communication.

bluetoothctl
A command-line Bluetooth controller program.

boolean
A common type in C++ programming. A boolean is a binary true or false value.

breakout board
Boards that break out the functionality of your Edison to a larger, easier-to-access module. The three main breakout boards for Edison are the Intel Arduino Breakout Board, Intel Mini Breakout Board, and SparkFun Base Block.

byte
A common type in C++ programming. A byte is an integer from 0 to 255.

cathode
The lower voltage side of a polarized electrical device.

char
A common type in C++ programming. A char is an integer from -128 to 127.

circuit diagram
A graphical representation of an electrical circuit.

computer vision
The gathering, processing, analyzing, and understanding of images and video data. Computer vision is a complex, widely studied, and growing field in the scientific world today.

debouncing
The process of removing fast fluctuations over a short period of time from a digital signal. Debouncing can be done in either hardware or software.

dependencies
Software packages or libraries that another software package requires to run.

directory
A folder in the file system of an operating system.

duty cycle
The percentage of time that a pulse-width modulated signal spends in the high state. See also Pulse Width Modulation (PWM).

embedded devices
Devices buried within the hardware of a system and which are often used to control the system in real time.

flags
Optional command-line tags that set the configuration for any specific command and are issued after the name of the command itself.

float
: A common type in C++ programming. A long is a decimal value from -3.4028235E38 to 3.4028235E38.

`for` *loop*
: A loop that steps through its associated code block so long as the `while` condition is true.

function
: In software, a function is a self-contained piece of code that runs independently of the rest of the program.

ground
: The return point where the current in closed-loop circuits often terminates.

home directory
: A file-system directory that containes files for a given user or a multi-user operating system. On Intel Edison, the path to this directory is */home/__username__/* be default.

I2C
: A protocol developed by Philips Semiconductors in the late 1970s to simplify the lines that travel among various circuit components. The I2C protocol reduces the number of communication wires to two by allowing multiple peripherals to communicate on the same lines.

input
: In hardware, this is a signal that transmits information into your Intel Edison from the outside world. The press of a button and measurements from a thermometer are good examples of inputs.

`int`
: A common type in C++ programming. An `int` is an integer value from -32768 to 32767. `int` is probably the most common type used in Arduino programming for beginners.

Intel Edison compute module
: The integrated hardware and software that comprises the Intel Edison in the absence of any breakout boards or peripherals.

Intel Mini Breakout Board

A breakout option for Intel Edison that gives solderable access to Intel Edison's I/Os. This breakout board is smaller than the Intel Arduino Breakout Board and is somewhat more difficult to work with. The Intel Mini Breakout Board, SparkFun Base Block, and Intel Arduino Breakout Board are the three main breakout boards for Intel Edison. See also Arduino Breakout Board, SparkFun Base Block.

interpreted language

A language that allows you to execute scripts directly through an interpreter instead of first compiling them into machine code.

level shifters

Hardware that converts voltage signals from one level to another, such as from 1.8V to 5V, or vice versa.

light-emitting diode (LED)

A semiconductor device that emits light when a current passes through it.

Linux

Linux is a free and open source operating system (OS) that runs on top of the Linux kernel.

load

Any part of an electrical circuit between the source and return. The load in an electrical circuit can be pretty much anything: lights, resistors, screens, speakers, and even the wire making the connections. See also Source, Return.

long

A common type in C++ programming. A long is an integer value from -2,147,483,648 to 2,147,483,647.

loop

A function that every Arduino sketch must contain. After the `setup` function completes, this function runs over and over again until power is removed from the device or a new sketch is loaded.

OpenCV

An open source computer-vision software platform.

output
: In hardware, this is a signal that is emitted by your Intel Edison, often to control attached devices. For instance, Edison might output voltage signals to make lights flicker or display images on a screen.

path
: The full, absolute location of a file or directory on an operating system's file system.

pip
: The Python package installer.

piping
: At the command line, piping takes the output of one command and supplies it as the input to another, using the | symbol as the pipe operator.

pull-down resistor
: A resistor connected to ground that holds a digital signal at 0V when nothing else is connected.

pulse-width modulation (PWM)
: The rapid oscillation of an output signal that is fast enough for connected devices to experience the average voltage instead of the rapidly oscillating high or low.

rails
: Columns on a breadboard that are electrically connected.

register
: An internal address or location of configuration, data, or other elements of an I2C device.

regular expressions
: A powerful template-matching system for words, numbers, and symbols.

return
: The spot at which a current terminates in a circuit, most frequently ground.

root mean square (RMS)
A numerical value calculated by squaring every data point in the sample, taking the mean of these values, and then taking the square root of that mean.

serial
A protocol for communication between electronic devices.

serial peripheral interface (SPI)
A standard protocol used to connect multiple peripherals to the same data lines. SPI operates using a slave-master architecture, meaning one device has full control over all peripherals. Each SPI device requires four connections, one of which is not shared with other SPI devices.

serial port profile (SPP)
A protocol that emulates a serial cable for wireless communication such as Bluetooth.

setup
A function that every Arduino sketch must contain. This function is run exactly once when the device is booted or the sketch is first loaded onto the device.

sketch
In the Arduino IDE, a sketch represents standalone code to control your Arduino or Intel Edison. Each sketch must contain a `setup` and a `loop` function.

source
The spot from which a current originates in a circuit, such as a battery or power supply. The source has a higher voltage than the return, or ground. See also Return.

SparkFun Base Block
A modular breakout option for Intel Edison. The Base Block breaks out just a few functions of the Intel Edison and is meant to stack with other SparkFun blocks for increased functionality. The SparkFun Base Block, Intel Mini Breakout Board, and Intel Arduino Breakout Board are the three main breakout boards for Intel Edison. See also Intel Mini Breakout, Arduino Breakout Board.

unsigned char
A common type in C++ programming. An unsigned char is an integer from 0 to 255.

unsigned int
A common type in C++ programming. An unsigned int is an integer value between 0 to 65535.

unsigned long
A common type in C++ programming. An unsigned long is an integer value from 0 to 4,294,967,295.

variable
In software, a reference to stored information and the name is typically chosen to represent the information it contains.

`while` *loop*
A loop that repeats its associated code block at each increment in the `for` statement.

word
A common type in C++ programming. A word is an integer value from 0 to 65535.

Yocto Linux
The version of Linux running on Intel Edison. Yocto Linux is designed specifically for embedded devices.

Index

Symbols

- && (logical and) operator, 74
- * (wildcard) character, 35
- += (compound assignment) operator, 77
- . (dot), . and .. notation in Linux, 30
- / command in Linux, 29
- /* and */ in C++ comments, 56
- // in C++ comments, 56
- : (colon) in Python code, 101
- ; (semicolon) in C++ code, 58
- > and >> (redirection) operators, 38
- { } (curly brackets), enclosing C++ code blocks, 55
- | (pipe) operator, 40, 175
- ~ (tilde), representing home directory, 28

A

- accelerometers, 51
  - I2C, 82
    - recreating example in Python, 113
  - use in wearable devices, 118
- Adafruit GFX library, 89
- Adafruit ILI9341 library, 89
- Adafruit MMA8451 library, 86
- Adafruit sensor library, 86
- Adafruit TFT Touch Screen Display, 88
- Advanced Packaging Tool (APT), 161
- alsa utilities for sound, 145
- ambient noise, adjusting for, 153
- analog input, 78-82
- analog output, 75-78
- analog-to-digital converter (ADC), 78, 171
- analog.py script, 112
- analogRead function, 79
- analogWrite function, 76, 81
- Android devices, 106
  - BlueTerm app, 109
- anode, 63, 171
- aplay command, 144, 146
- apt-get utility, 161
- Arduino, 49-96
  - analog input, 78-82
  - analog output, 75-78
  - blink program (example), 57
    - blink circuit, 61
    - expanding, 59
  - defined, 49, 171
  - digital input, adding a button, 68
    - monitoring button state with
      - serial console, 70
  - I2C accelerometer, 82
  - materials list, 50
  - sketches and functions, 55
  - SPI screen, 87
  - troubleshooting sketches and libraries moved to Edison, 95
  - using Intel Edison with, ix
- Arduino Breakout Board, 2
  - components, 3
  - setup and configuration, 11
- Arduino IDE, 14, 52
  - C++, Linux and, 92
  - configuring to communicate with Edison, 54
  - installing, 52
  - navigating, 52
  - Serial Monitor, 71
- Arduino-style starter packs, 50
- arecord command, 146

arithmetic operators in compound
assignment operators, 77
audio, 143
(see also sound, exploring)
recording with Python, 148
basic recording, 148
audioloop library, 150
avrlibc library, 171

B

barrel jack (Arduino breakout
board), 4
battery packs, 147
baud rate, 70, 171
problems caused by, 71
BGR (blue, green, and red) values,
128
BLE (Bluetooth Low Energy) beacon (see Bluetooth)
blink program (example), 57
blink circuit, 61
expanding, 59
in Python, 103
blinkIt function (example), 60
BlueTerm app, 109
Bluetooth
BLE beacon, turning Edison into,
120-123
enabling on Edison, 107, 122
pairing Bluetooth devices, 107
using to toggle LED from smartphone, 106
bluetoothctl program, 107, 171
Boards Manager, Arduino IDE, 55
booleans, 56, 171
Bourne shell, 29
writing series of commands to
file and executing them
sequentially, 41
breadboards, 50
solder-less, electrical connections inside of, 62
breakout boards, 2, 126
analog input and, 79
Arduino breakout board, 3
defined, 171
other breakout options, 6
brightness variable, 77
potentiometer example, 81
adjusting LED brightness, 81
buttonPin variable, 68
buttons
adding push button to LED circuit, 68
toggling the LED, 73
Arduino breakout board, 4
button-controlled blink in
Python, 106
for Arduino breadboard, 51
buttonState variable, 68
byte type, 172

C

C language, 163
C++, 162
and Arduino IDE, 92
comments in, 56
data types, 55
defining functions, 55
semicolon (;) in code, 58
Camera object, 138
cap.read function, 134
cap.release function, 135
cat command, 39
cathode, 63, 172
cd (change directory) command,
29
char type, 56, 172
chmod command, 37
chown command, 36
circuit diagrams, 61, 172
for button example, 72
circuits
common elements and representations in circuit diagrams, 61
defined, 61
integrating a button into blink
circuit, 106
layout for blink example, 63
layout for button example, 69

layout for potentiometer example, 80
powering down before changing connections, 62
class function (Python), 138
clear command, 34
codecs (video), 137
command line, 28, 29
commands
flags, 30
piping output into another command, 40
comments in C++, 56
communications device, Intel Edison as, vii
communications port, selecting for Arduino IDE, 54
compiling C code, 163
compoound assignment operators, 77
compressed files, 43
computer vision, 125-142, 172
materials list, 126
OpenCV, 127
extracting colored objects, 127-128
face detection, 130-133
viewing images, 128
performing face detection using live camera, 133-141
computers
connecting Arduino breakout to, 12
Intel Edison as a computer, vii
configure_edison --setup command, 20
continue statement (Python), 137
cp (copy) command, 32
Ctrl+A command-line shortcut, 29
Ctrl+C command-line shortcut, 29
Ctrl+D command-line shortcut, 38
Ctrl+E command-line shortcut, 29
curl command, 42
custom library for ILI9341, 89
cv2.imwrite function, 135
cv2.VideoCapture function, 134
cv2.VideoWriter function, 136
CyberDuck, 43

D

date command, 34
Debian Linux on Edison, 159
debouncing, 75, 172
debugger messages in Edison console, 139
def function (Python), 138
delay function
for analog output, 77
for serial console, 71
delayMicroseconds function, 58
delayTime variable, declaring, 59
dependencies, 172
package, 44
df command, 34
digital input
using a push button, 68
monitoring button state with serial console, 70
toggling the LED, 73
digital signal processing, 156
digital signals, voltage, 10
digitalRead function, 73
digitalWrite function, 58
diodes, 66
directories, 26, 172
changing in Linux relative to current directory, 30
creating with mkdir command, 31
important paths on Intel Edison, 27
-r (recursive) flag and, 33
dmesg command, 40, 134
duty cycle, 76, 172

E

echo command, 38
Eclipse IDE, 14
edge detection, 140
Edison Forums, 95
Edison Troubleshooting Guide, 95
elif statement, 156
else statement, 73

embedded devices, 26, 172
espeak utility, 146
except block (try-except in Python), 115
execute permission, 36
external USB port (Arduino breakout board), 4

F

face detection, 130-133
  performing with live camera, 133-141
FacialRecognition.py script, 130
female 70-pin connector (Arduino breakout board), 4
file transfers, 129
files
  copying with cp command, 32
  deleting with rm command, 33
  moving with mv command, 32
  renaming using mv command, 32
  transfer programs for, 43
filesystems
  comparison for Linux and Windows, 26
  Edison filesystem, 26-29
  navigating, basic Linux commands for, 29
FileZilla, 43
find command, 34
flags (command), 30, 172
  -r (recursive) flag, 33
  different versions of, 31
flashing Edison with latest build, 14
  reconfiguring Edison after, 23
Flask library, 137
float type, 56, 173
for loops, 101, 173
FTP (File Transfer Protocol), 43
functions, 173
  defining in C++, 55
  preprogrammed into Arduino, 58
  Python, 101
  required, in Arduino sketches, 55

G

getting help, xiii
git
  git clone versus wget command, 116
  installing, 45
Google Speech Recognition API, 152
grep command, 34
  piping dmesg command output to, 134
  piping output of ps command into, 40
  piping top command output to, 112
ground, 61, 173

H

hardware development platform, Intel Edison as, vii
head command, 39
headphone jacks, 144
headsets, 144
  connecting to Edison, 144
"Hello World" program in Arduino IDE, 89
"Hello, World" program in Python, 98
help statement (Python), 100
hexadecimal numbers, 84
hidden files and directories in Linux, 30
home directory, 28, 173
  switching back to, commands for, 29

I

I2C accelerometers, 82
  programming by Arduino community, 86
  recreating example in Python, 113

wiring for, 82
I2C protocol, 173
ibeacon library, 120
IDEs (integrated development environments), 49
Arduino IDE, 52
if statement, 73, 156
checking for multiple conditions, 74
performing checks in I2C accelerometer example, 84
if-else statement, 73
ILI9341 controller, 89, 119
custom library for, 89, 154
ImageDraw object (Python), 119
images
drawing and manipulating in Python, 118
drawing to screen in Arduino IDE, 92
extracting a blue frog (example), 128
reading in OpenCV and converting to grayscale, 132
viewing using file transfer, 129
viewing with SPI screen, 128
import statement (Python), 99
imwrite function, 135
indentation in Python code, 101
input, 173
input and output (I/O) pins, Arduino breakout board, 5
installers, stand-alone, 14
int type, 56, 173
integrated development environments (see IDEs)
Intel Edison
about, vii
assembly onto Arduino Breakout Board, 11
Bluetooth and, 106
connecting to, 16
configuring Edison and getting online, 20
from Linux, 19
from Mac, 19
from Windows, 16
logging in, 20
different Linux flavors on, 159
installing on your computer, 13
Mac and Windows, 14
limitations of, vii
projects, x
shutting down, 23
troubleshooting, 23
turning into BLE beacon, 120
uses of, viii
Intel Edison compute module, 1, 173
Intel Galileo, 104
shields for, 95
XDK for, 164
Intel Mini Breakout Board, 6, 126, 174
Intel Quark microcontroller, 52
Intel XDK IoT Edition, 14, 164
interactive shell in Python, 98
opening, 98
tab completion support, 99
Internet
connecting to, from Edison, 20
Intel Edison and, 93
interacting with, 41-44
Internet of Things (IoT), vii, 164
interpreted languages, 97, 174
iOS, 98
Bluetooth pairing, 106
IP address
for Edison, 93
getting for Edison, 43

J

JavaScript, 161

K

keyboard shortcuts, 29, 54
kill command, 112
using PIDs with, 40
Kirchhoff's second law, 66

L

LastFM, free download collection, 147
ledPin variable, 73
  declaring, 59
ledPinAnalog variable, 81
LEDs, 50, 174
  controlling output using Edison, in blink program, 57
  in circuits, 61
  power LED indicator, Arduino breakout board, 5
  toggling using Bluetooth connection on smartphone, 106
ledState variable, 73
less command, 39
level shifters, 10, 174
libraries
  for I2C accelerometers, 86
  for SPI screen example, 89
  installing in Python, 102
Linux, 25-47, 174
  basic commands, 29-38
  C++ and the Arduino IDE, 92
  connecting to Intel Edison, 19
  defined, 25
  different flavors, Intel Edison and, 159
  Edison filesystem, 26-29
    accounts, permissions, and ownership, 35
  ibeacon library, 120
  installing Intel Edison, 15
  installing packages in Yocto Linux, 44
  interacting with the Internet, 41-44
  Linux-compatible USB headset, 144
  scripting and more advanced commands, 38-41
  text editors, 46
  Yocto Linux for Intel Edison, 26
listen command, 153
load, 61, 174
Locate Beacon app, 122
logging in to Intel Edison, 20
logging, turning off in Edison, 46
Logitech ClearChat Comfort/USB Headset H390, 144
long type, 56, 174
loop function, 55
  calling blinkIt function (example), 60
  for analog output, 77
  for SPI screen example, 91
  in blink program, 57
  potentiometer example, 81
  printing out button state to serial console, 70
  writing code fo monitor state of button, 68
looping back (SPP-loopback.py), 109
  replacing loopback with LED controls, 111
loops, 101, 174
ls command, 30
  flags, 30
lsusb command, 133
Lynx command-line web browser, 42

M

MAC address, 108
Mac computers
  communications port for Arduino IDE, 54
  connecting to Intel Edison, 19
  installing Intel Edison, 14
main function, 163
major identifier, 121
male-to-male breadboard wires, 50
map function, 81
materials, 167-169
microcontrollers, 49
microphone jacks, 144
microphones
  3.5 mm, 144
  ambient noise and speech volume, 153
microSD Block, 147

MicroSD cards, 130
MicroSD slot (Arduino breakout board), 4
microUSB cables, 3
microUSB port, 5
  connecting to your computer, 13
minor identifier, 121
MISO pins, 88
mkdir command, 31
MMA8451 library, 86, 115
more command, 39
MOSI pins, 88
mp3 player, 146
mpg123 library, 147
mraa library
  installing, 45
  odd GPIO pin numbering schemes, 105
  using in Python potentiometer example, 113
mv (move) command, 32

N

nano (text editor), 46
natural language processing, 156
Node.js, 161
Node.js web server, 21
NumPy package, installing, 115, 127

O

Ohm's Law, 65
OLED Block, 147
On-The-Go (OTG) microUSB port, 5
  on Intel Mini Breakout and SparkFun Base Block, 7
OpenCV, 127-133
  extracting colored objects, 127
  face detection, 130
  viewing images, 128
opkg utility, 44, 160
  pip package manager versus, 103
  updating to avoid errors, 145
output, 175
ownership of files, 36

P

packages
  installing in Yocto Linux, 44
  installing, using opkg or pip, 103
  viewing all available packages, 46
  Yocto Linux for Intel Edison, 160
passwd command, 37
paths, 26, 175
  important paths on Intel Edison, 27
permissions, 35
  read, write, and execute, 36
Phone Flash Tool, 14
photos, snapping with a webcam, 134
PIDs (process IDs), 40
pin-mapping and mraa documentation, 105
ping command, 41
pinMode function, 57
  button pin as input, 68
pip (package manager), 102
piping, 40, 175
playlist, creating, 147
potentiometers, 51, 61, 79
  potentiometer example in Python, 112
potPinAnalog variable, 81
potValue variable, 81
power LED indicator (Arduino breakout board), 5
power mode microswitch (Arduino breakout board), 5
power supplies, 126
programming languages, 161
  C and C++, 162
  Node.js, 161
ps command, 40
pull-down resistors, 72, 175
pull-up resistors, 72
pulse width modulation (PWM), 75, 113, 175
pushbuttons, 61
PuTTY

configuration for connecting to Intel Edison, 18
configuration for wireless connection to Intel Edison, 21
downloading and installing, 17
pwd (present working directory) command, 30
PWM-enabled pins, 76
.py file extension, 100
pyaudio library, 148
pyserial library, 103
Python, 97-124
advantages as first programming language, 97
BLE beacon, turning Edison into, 120
Edison side, 120
smartphone side, 122
blink program (example), 103
button-controlled blink (example), 106
computer vision and OpenCV resources, 142
creating standalone scripts, 100
functions and loops, 101
"Hello, World" program, 98
I2C accelerometer example, 113
installing the dependencies, 115
using the MMA library, 115
installing libraries, 102
mathematical computations, 99
Node.js versus, 161
OpenCV in, 127-133
extracting colored objects, 127
facial recognition, 130
performing face detection using live camera
recording video, 135
snapping photos, 134
video streaming, 137
potentiometer example, 112
recording audio with, 148
basic recording, 148
thresholding, 150
resources for further learning, 123
speech recognition, 152
using to control devices, 154
SPI screen example, 118
starting scripts at bootup, 105
versions, 99

R

-r (recursive) flag, 33
rails (breadboard), 62, 175
read function, 134
read permission, 36
redirecting command-line output, 38
registers (device), 83, 175
writing WHOAMI register to I2C accelerometer, 84
regular expressions, 34, 175
release function, 135
renaming files with mv, 32
resistive touch breakout, 51
resistive touch shields, 51
resistors, 51, 61
calculating for LED circuits, 67
pull-down and pull-up, 72
return (or ground), 61, 175
return type, 55
rfkill command, 107
rm (remove) command, 33
root mean square (RMS), 150, 176
root user, 28
file ownership, 36

S

SCL pins, 82
scp command, 43
scp utility, 130
screen program
connecting to Edison on Linux, 16, 19
important commands, 20
using to connect to Edison on Mac computer, 19
scripts
Linux, shell scripts in, 41

Python
creating standalone script, 100
running as executable, 121
running at startup, 105
SDA pins, 82
serial console, 70
viewing values read from any analog sensor, 81
serial console USB port (Arduino breakout board, 5
serial monitor (Arduino IDE), 71
Serial Peripheral Interface (SPI), 87, 176
(see also SPI screens)
serial port profile (SPP), 109, 176
serial protocol, 176
Serial.begin function, 70, 84
Serial.println function, 70
setup function, 55, 176
enabling serial console, 70
for I2C accelerometer example, 84
for SPI screen example, 90
in blink program, 57
initializing button pin as input, 68
SFTP (SSH file transfer protocol), 43
sftp command, 43
sftp utility, 130
sh command, 41
.sh file extension, 41
shells
Bourne shell, 29
opening interactive shell in Python, 98
shields, Galileo-compatible, 95
shutdown, 23
sketches, 52, 176
and functions, 55
digital input, adding a button, 68
for I2C accelerometer example, 85
saving, 59
toggling the LED, 74
translation into Python, 106
smartphones, 98
BLE beacon example, 122
Bluetooth pairing with Intel Edison, 108
enabling Bluetooth discovery, 108
exchanging information with Intel Edison, 109
using Bluetooth connection to toggle LED, 106
soldering irons, 51
sound, exploring, 143-157
connecting a headset, 144
creating makeshift mp3 player, 146
materials list, 143
recording and playing sound, 145
recording audio with Python, 148
thesholding, 150
speech recognition, 152
using speech recognition to control devices, 154
sound_recorder.py script, 148
source, 61, 176
SparkFun Base and Battery Blocks, 147
SparkFun Base Block, 6, 126, 176
speech recognition, 152
using to control devices, 154
SpeechRecognition library, 152, 154
SPI screens, 87
controlling using speech recognition, 154
integrated electronics, 89
recreating example in Python, 118
SPI-driven displays, 51
SPP (Serial Port Profile)-
loopback.py, 109
modified version, 110
SS pins, 88
starter packs (Arduino), 50
stepSize variable, 77
streaming video, 137

processed, 140
system function, 92

T

tail command, 39
piping dmesg command output to, 40
tar command, 43
terminal programs, 17
text editors, 46
thresholding audio, 150, 153
top command, 112
Torvalds, Linus, 25
touch command, 31
transmission (tx) power, 122
troubleshooting, 23
Arduino sketches and examples moved to Edison, 95
try-except statement (Python), 115
types (C++), 55

U

Ubilinux on Edison, 159
umount command, 130
unsigned char, 56, 177
unsigned int, 55, 177
unsigned long, 56, 177
urllib library, 127
USB buses, getting information about, 133
USB devices
USB headset, 144
USB to 3.5 mm jack converters, 144
USB hub, 134
USB port, external, for Arduino breakout board, 4
useradd command, 37
UUIDs (universally unique identifiers), 121
UVC-compatible webcams, 126

V

variables
defined, 177
defining in Arduino, 59
vi (text editor), 46
video
procesed streaming video, 140
recording from a webcam, 135
streaming, 137
video codecs, 137
VideoCapture command, 134
VideoWriter function, 136
vision (see computer vision)
VLC Player, 137
voltage
digital signals, 10
Ohm's Law, 65
reduction by resistors and diodes, 66
specifying for analog output, 76
Vout signal, 79

W

wave files, 145
wearable electronics, 118
web content, serving through Arduino IDE, 93
web servers
node.js web server in Edison, 21
video streaming server, 137, 140
webcams, 133-141
recording video, 135
snapping photos, 134
UVC-compatible, 126
wget command, 42
while loops, 102, 177
while loops (Python), 110
whitespace in Python code, 101
WHOAMI register, 84
Windows systems
communications port for Arduino IDE, 54
connecting to Intel Edison, 16
installing Intel Edison, 14
Wire library, 83
Wire.available function, 84
Wire.begin function, 84
Wire.beginTransmission function, 84

wireless connections to Edison, 21
wireless file transfer, 130
wiring, I2C accelerometer example, 82
word type, 56, 177
write permission, 36

X

XDK IoT Edition, 164

Y

Yocto Linux, 13, 25, 26, 177
(see also Linux)
installing packages, 44
switching to Ubilinux from, 160

About the Author

Stephanie Moyerman is a research scientist in the New Devices Group at Intel. Her work focuses on innovation and proof of concept demonstrations for wearable technology. She graduated with a Ph.D in astrophysics from the University of California, San Diego in 2013. Before that, she attended Harvey Mudd College and received dual B.S. degrees in math and physics.

Outside of work, Stephanie's favorite activity is judo. She's been a junior and collegiate national champion and has been ranked as high as #5 in the United States. She also enjoys surfing, wakeboarding, running, glass blowing, spoiling her dog, and—of course—making.

Colophon

The cover photo was taken by Brian Jepson. The cover fonts are Benton Sans Bold, Benton Sans Light, and Soho Pro. The text font is Benton Sans, the display font is Serifa, and the code font is TheSansMono Condensed Regular.

CPSIA information can be obtained at www.ICGtesting.com
Printed in the USA
BVOW11s0030281115

428270BV00003B/3/P